Ancient Light from this Mountain

Philip Ernest Williams

Revision 2, August 2022

ISBN 978-0-9793102-3-2

Library of Congress Control Number: 2022913955

Published by:

Christian Leaders & Scholars Press

5155 Swearngan Road, Charlotte, NC 28216

Cover Photo: Mount Ararat from Dogubayazit, Turkey

To the late ***Dr. Norman Geisler (1932-2019)***

who so eagerly desired that I complete this book

concerning the only view of origins that

he ever endorsed with his name and scholarship

The author, Norm Geisler, Ahmet Ertugrul

PREFACE

The author wrote this monograph for the *International Conference on the Archaeological Traces of Noah's Flood on Mount Ararat* which took place May 5-7, 2022 near the town of Dogubayazit, Turkey. This symposium was held at *Noah's Village* located on the road that traverses the south side of Agri Dagh, a mountain that is better known in most of the world as Mount Ararat.

This symposium of more than fifty archaeologists and other scientists and scholars from Turkey, Europe, the United States, and various nations of the region together with officials from the Governments of the Republic of Turkey and provinces near this mountain was sponsored by *Andrews University* of Berrien Springs, Michigan and by the North Carolina-based *Mount Ararat Discovery Foundation*. The latter is a 501(c)3 non-profit devoted to the preservation and authentication of artifacts found in the region of Mount Ararat and to the publication of scientific studies pertaining to these artifacts in peer-reviewed journals, all in cooperation with the Governments of the Republic of Turkey and its provinces and with the natives of this mountain and region. The conference was hosted by the *Dogubayazit Faith Tourism, Development, and Research Association* chaired by Ahmet Ertugrul.

This monograph, a summary of the author's more than thirty years of extensive research, was written to assist the archaeological investigation of artifacts and what are reported to be an ancient wooden structure buried near the top of this mountain which the natives have always believed to be Noah's Ark. The author serves as Scientific Advisor to the *Mount Ararat Discovery Foundation.*

References to so much scientific history that is hardly mentioned in popular scientific writings, and which differs from popular scientific histories may challenge many readers due to the vast amount of information summarized in this short treatise. Even so, I trust the reader will benefit from learning just how and why our current science came to understand the things that we are taught about the history of ancient man and the recent history of the earth.

Likely, the reader will be astonished to learn the strange and interesting way that the current historical sciences came about, but no less astonished by how simply the same evidence appears in the light of the narrative found in our sacred scriptures. The reader will understand why the late Dr. Norm Geisler believed the discoveries on this famous mountain understood in the light of the new understanding presented in this book could change our world.

Philip Ernest Williams Charlotte, NC

CONTENTS

Chapter 1

INTRODUCTION

Many are familiar with the numerous reports of the remains of Noah's Ark still resting on the top of this mountain by which we are gathered. Some of these reports date from before the time of Jesus. Josephus, the first century Jewish historian, writes that Berossus the Chaldean (3rd century BC) tells us that the Ark 'was preserved when it was brought to the highest part of the mountains of Armenia,' as this region was known in the first century.[1] That would require the Ark to have rested near the top of this mountain which we may be able to view later today as the covering clouds lift from the mountain.

I have however never been interested in searching for Noah's Ark. Nor was I much interested in the many claims for the discovery of the Ark which I formerly supposed could not have survived for thousands of years. That was before I learned that organic remains could survive for so long if they have been constantly frozen as is the case of 5,000-year old Otzi the Iceman who was found under a glacier in the Alps back in the nineties. But my interest in these

---

[1] Josephus, Whiston translation, *Against Alpion*, Book I, Chapter 19, passage 130

reports, ancient and modern, was because I had traced a secondary worldwide dispersion of mankind from the Ancient Near East which begins near this mountain. I was interested in knowing whether there are artifacts in the ancient wooden structure reported to be on this mountain that might be related to those of the widespread dispersion from near this mountain that archaeologists identified many years ago (see map below).[2]

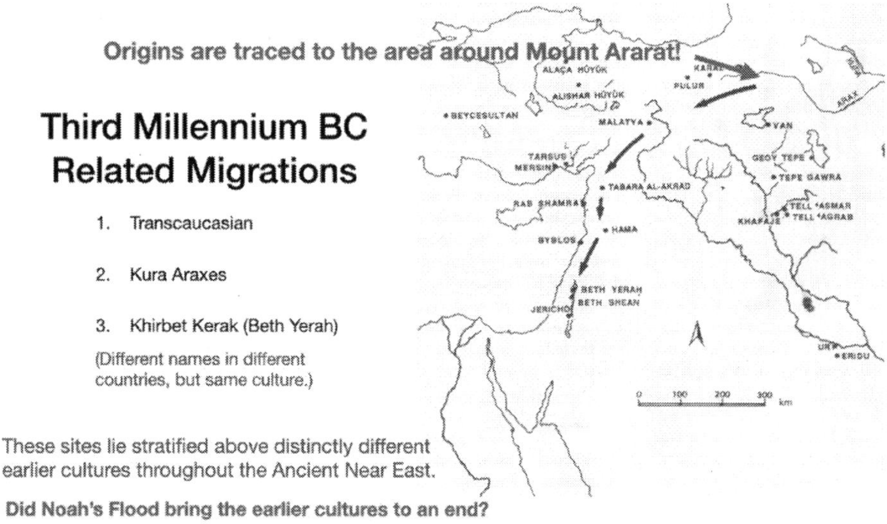

The map above shows only the migration arm which spreads into the southern Levant. I will say more about the wider dispersion below. But should the wooden structure reported to be buried near the top of this mountain be the source of the archaeological culture

---

[2] Map and migration path from Ammon Ben-Tor, *The Archaeology of Ancient Israel,* (Yale University Press, 1994), p 123. Text at left is from this author.

spreading from this mountain, it will point to this region as the ancient homeland of the people of every nation who scattered across the earth carrying with them memories of a great world-destroying Flood.

I am grateful that the Government of Turkey is at last able to allow archaeological excavations on this frontier mountain which for many centuries has been the center of much trouble and danger. This mountain and region may soon become the center of worldwide scientific and historical investigations, helping bring peace to the nations and much attention to this region as the world discovers the land near where we are meeting is the ancestral homeland of us all.

The reason for my presentation today concerns the *archaeological* identification of artifacts that may be found inside the ancient wooden structure. Should they connect to the archaeological culture that spreads from this region (see map at left), the structure belongs to what archaeologists currently understand as the prehistoric era. It will date to the Chalcolithic Age when both copper and stone tools were being used. That is a millennium earlier than Archbishop James Ussher's famous 2349BC dating of the Flood. I believe the error lies not so much with Ussher but with the current prehistoric chronology. Whatever the case, we may have in this project the means of determining just where the dating error does lie.

Determining this requires learning the strange reasoning behind the establishment of archaeology's prehistoric chronology, the full story of which no historian of ideas has previously researched and published. Thus, it was necessary that I investigate just how archaeologists came to understand and date man's prehistory. Here

I will summarize the story of how a few thousand years of mankind's earliest history got transformed into a few hundred thousands of years, leading to the dating problem that I just mentioned. It is also how the light of the sacred scriptures got replaced with speculative theories based entirely on naturalistic causes.

A paradigm shift for the historical sciences

I am sympathetic to how shocking this must seem to those who are well-trained in the historical sciences. It is foolishness because contrary to well-known facts. I could hardly expect my fellow scientists to be interested were it not for reports of an ancient wooden structure resting exactly at the one tiny spot on our large earth's surface that suits the landing place of Noah's Ark as told in the sacred scriptures. It must necessarily have rested near the top of the tallest(!) of the mountains of Ararat if the tops of the surrounding mountains weren't visible when the Ark came to rest.[3]

Alfred North Whitehead, who as President of the Royal Society defended Einstein's new theory of relativity, reminds us that there are no uninterpreted facts. All observations must be interpreted according to some theory or narrative. Nowadays, such a light is called a paradigm. Thomas Kuhn, the philosopher of science who popularized the concept of paradigm, explains the difference between normal and revolutionary science. So implicit is our trust in what we are taught to believe that during times of normal science a contrary paradigm will seem foolish.

---

[3] c.f. Genesis 8:5

The paradigm that we so implicitly trust that we may not even be conscious of using is the light by which we ordinarily see and judge things. Because such understandings are believed *a priori*, they cannot be used to judge a different paradigm for understanding evidence. There is however a way of judging between paradigms. Another philosopher of science Karl Popper pointed out that good paradigms make specific and therefore testable predictions.

For example, in the light of the evolutionary paradigm, we should have found the still missing links between species. In the light of young earth Creation Science, there should be massive evidence of civilizations drowned by the worldwide Flood in the third millennium before Christ. Unfortunately, the failure to find such evidence can either be ignored or else explained away by *ad hoc* theories. Evolutionists tell us that not even one of the many evolutionary links that must exist between species was preserved in the fossil record. Similarly, the Creationists tell us that the Flood itself wiped away all the evidence of the many people together with their dwellings and cities who lived before or who died in the Flood.

Though I will put forth here manifest evidence of a worldwide Flood and the recent appearance of man, I am not a young earth Creationist. Many young earth Creationists are troubled by reports of the remains of Noah's Ark being found on this mountain which sits atop of what they tell us is the geological evidence of the Flood. They believe this mountain must be a post-Flood volcano which didn't exist when the Ark rested on the mountains of Ararat. Some young earth Creationists have a different view, but my old earth science is so different from *Creation Science* that I coined for it the term *Scripture Science*. Scripture Science differs from both mainstream science and Creation Science in that its light is neither a scientific nor a theological system but a simple historical narrative.

Whatever one must or chooses to believe, some type of paradigm is unavoidable due to it being the light by which we turn observations into what we understand to be the facts. To judge paradigms, whether it be the uniform laws of the current historical sciences, the young earth view of the Creationists, or the one I use here we need first understand the circumstances that led to their establishment. Then we should compare what we see in the light of these various paradigms. Just as focusing the Hubble telescope's lens gave a clearer picture of the beginnings of our physical universe, replacing the paradigm of the current historical sciences with a true one ought to give a clearer and simpler understanding of mankind's most ancient history. It should allow us to predict many things that we now understand as isolated facts.

The paradigm I will use here however is not a new light but an old lamp which our modern world has discarded due to it being seen as too ancient. But having used it to understand the evidence of the current historical sciences, I found it a powerful light for simplifying and clarifying our understanding of origins.

Chapter 2

THE ARCHAEOLOGICAL EVIDENCE OF NOAH'S FLOOD

In 2011, I published the results of my twenty-year search for archaeological evidence of Noah's Flood as should exist in the light of the narrative in the book of Genesis.[4] I aimed to find out whether such evidence even exists and if so, what might be the Flood's geographical extent. According to the methodology I employed, the four criteria I decided should serve as definite archaeological evidence of the Flood were:

(1) the sudden disappearance of the earth's population in the third millennium before Christ,

(2) followed by a secondary worldwide dispersion of mankind from the Ancient Near East,

(3) a clear break in material culture every place where appears this secondary dispersion, and

(4) evidence that the sudden ending of the world's population was due to drowning.

---

[4] Philip Ernest Williams, *The Archaeological Evidence of Noah's Flood (2011, CLS Press)*. Hereafter, I will refer to the book simply as the *Archaeological Evidence.*

I didn't initially look for evidence of drowning until I recognized it in the archaeological sites marking the world's first civilizations. But those include most prehistoric sites on every continent. After several years of unfruitful searching, to my great surprise I accidently happened upon the very kind of evidence for which I had been searching. Interestingly, I first discovered the evidence in Palestine, the land of ancient Israel. At the end of the Early Bronze Age, the population of this land disappears. I didn't make this discovery. The archaeologists had already noted it. To quote a prominent Israeli archaeologist:

A phenomenon so central in Palestinian history as that of the formation of urban society, with all its consequences for the material and spiritual culture of the country dissolves before our very eyes [Ammon Ben-Tor, *The Archaeology of Ancient Israel,* (Yale University Press, 1994), p 123]

The solvent for the Early Bronze Age cities was water. The Early Bronze Age ended about 2300 BC, very near the date of the Flood as can be calculated from the Bible. The first civilization to appear after this hiatus in population are the Canaanite city states, also mentioned in the Bible as the first civilization in this land following the Flood. These first Canaanites settlements, which include Ebla in Syria, mark the Middle Bronze Age in the Levant. While Egypt's Old Kingdom is presumed to date to the Early Bronze Age, the many scarabs of Old Kingdom Pharaohs found in the Levant are all from Middle Bronze Age sites. These correspond to the Sumerian and Akkadian periods in Mesopotamia, the Sumerian civilization possessing the only language that transcended this great interruption.

The first civilization after the Flood

This Middle Bronze Age begins from the migration shown on the map of page 2 above. The origins of this spreading migration should be some place close to where we are meeting. Depending on the countries where it is found, it is variously called *Transcaucasian, Kura-Araxes,* or *Khirbet Kerak.* Transcaucasia is the region straddling the South Caucasus mountains and where run the Kura and Araxes Rivers that are just north of this mountain. Archaeologists who study this culture tell us:

Kura-Araxes traditions apparently emerged during a rapid burst of cultural evolution, the early stages of which remain undocumented or unrecognized. [Peter Glumac and David Anthony, 'Culture and Environment in the Prehistoric Caucasus: The Neolithic through the Early Bronze Age,' published in *Chronologies in Old World Archaeology, Third Edition, Vol I* (1992, University of Chicago Press), edited by Robert W. Ehrich, p.203]

We may soon be able to document the amazing early stages of this culture should it be found to begin in the ancient wooden structure reported to be buried near the top of this mountain.

Khirbet Kerak, the third term used for this culture, is the Palestinian name for the ancient ruins found near where the Sea of Galilee empties into the Jordan River. It is the place where the eminent biblical archaeologist William F. Albright first identified this culture, which also defines the Early Bronze III (EBIII) era in the Ancient Near East. Notwithstanding how Albright named it, the Early Bronze III era is the true beginning of the Middle Bronze Age. Because the Early Bronze Age cities had just came to an end, Albright supposed these new people were invaders. But in contrast to the violence besetting the Early Bronze Age cities, these newcomers were peaceful. Such a peaceful culture could not have spread during a time when the violent civilizations of the thick-walled Early Bronze Age cities existed. Those spreading from these mountains of Ararat did often settle in or near the silt covered ruins of the Early Bronze Age cities which were locations central to travel or else had good water sources, rich land, or defensible sites.

The EBIII culture is identified by its distinctive red-black burnished ceramics which obtain their red coloration due to being baked in an open fire. This first culture after the Flood which we might suppose as being that of Noah's family began in these ancient mountains of Ararat which the Bible mentions as the landing place of Noah's Ark. Most interesting, the settlements of this first culture are found in the highlands as if the lower elevations are still covered by water![5] The thin-walled, key-hole shaped building foundations associated with these sites are suitable either for tents or wattle-and-daub housing. The latter may have developed by turning tents into permanent residences as we might expect of migrating populations. Fireplaces are found in the center of these houses where likely was made the red-black burnished pottery.

We are fortunate to have working on this project Dr. Stephen Pfann, who pioneered the use of neutron activation analysis using this same Khirbet Kerak pottery. Working with Charles Burney of the British Institute of Archaeology in Ankara, Pfann discovered that the pottery was both imported and locally made. That points to the

---

[5] Mitchel S. Rothman, 'KURA ARAXES CULTURE AREAS AND THE LATE 4TH AND EARLY 3RD MILLENIA BC POTTERY FROM VELI SEVIN'S SURVEYS IN MALATYA AND ELAZIG, TUKEY,' *ORIGINI* XXXVI 2014, p. 39

pottery spreading by migrations rather than trade, just what we should expect if this pottery was spread by families migrating from the Ark.[6]

There is another culture that extends over much of the same expansive region of the Ancient Near East as EBIII. It is associated with a city in ancient Mesopotamia called Uruk, but which the Bible refers to as Erech, one of the cities that Nimrod established in Shinar.[7] The name *Shinar* means 'the land of Sumer and Akkad.' The archaeological culture that spreads to or from this city which suddenly became the world's largest is called *Late Uruk.* Cylinder seals and bevel-rimmed bowls mark the spread of *Late Uruk.* Of the many cultures in the Ancient Near East, *Late Uruk* is by far the most widespread. These are just some of many reasons to suppose that the city of Uruk was also the site of the Tower of Babel. The bevel-rimmed bowls and the wooden-mold used for making them may be among the artifacts reported to be found in the ancient wooden structure that so interests us. So might the tree-of-life motif that is so widespread in the Mesopotamian culture of this era if not also some form of proto-Sumerian texts. If this is confirmed, it will transform our understanding of ancient history.

Late Uruk is currently dated earlier than the EBIII. This is because it was presumed to have immediately followed the earlier Uruk phases, which however were ended by the Flood. The bevel-

---

[6] Stephen J. Pfann, "Khirbet Kerak," in *Anchor Bible Dictionary* edited by Noel David Freedman (Yale University Press, 1992)

[7] Genesis 10:10

rimmed bowls, which were produced from a wooden mold, may have also served as a standard measure for trading grains and other produce, while the cylinder seals served as a mark of ownership. Likely, the cylinder seal was invented at Uruk where cuneiform writing was also invented. Some of these writings tell us that the origin of the Sumerians (a name which derives from Noah's son Shem) was in the land of Aratta, near the origins of the Tigris and Euphrates Rivers just south of this mountain. Arrata was obviously the land of ancient Ararat where later existed the Kingdom of Urartu. These Late Uruk markers may be spreading with the Khirbet Kerak, Kura-Araxes, and Transcaucasian EBIII culture beginning from inside the ancient wooden structure on this mountain.

The material culture of archaeological sites of the Middle Bronze Age in the Ancient Near East has numerous characteristics which spread across the earth along the migration routes on the map from the end sheets of my 2011 book *The Archaeological Evidence of Noah's Flood* (see map at right). Such intercontinental migrations as shown on the map will be unfamiliar to most readers. This is due to the cessation of the study of migrations and cultural diffusion that has informed archaeology since the sixties. Archaeologists no longer study worldwide migrations as shown on this map, reckoning that ancient agriculture and civilization developed independently in every land and continent. But this situation will soon change.

From the Middle Bronze Age of the Ancient Near East come the earliest historical cultures of every continent. The rich material culture carried by the migrants vary but they include spindle whorls and the heddle loom; waddle-and-daub huts and multi-generational houses with interior courtyards (*megarons*); the lost-wax form of bronze casting; the bow-and-arrow; bark paper and the

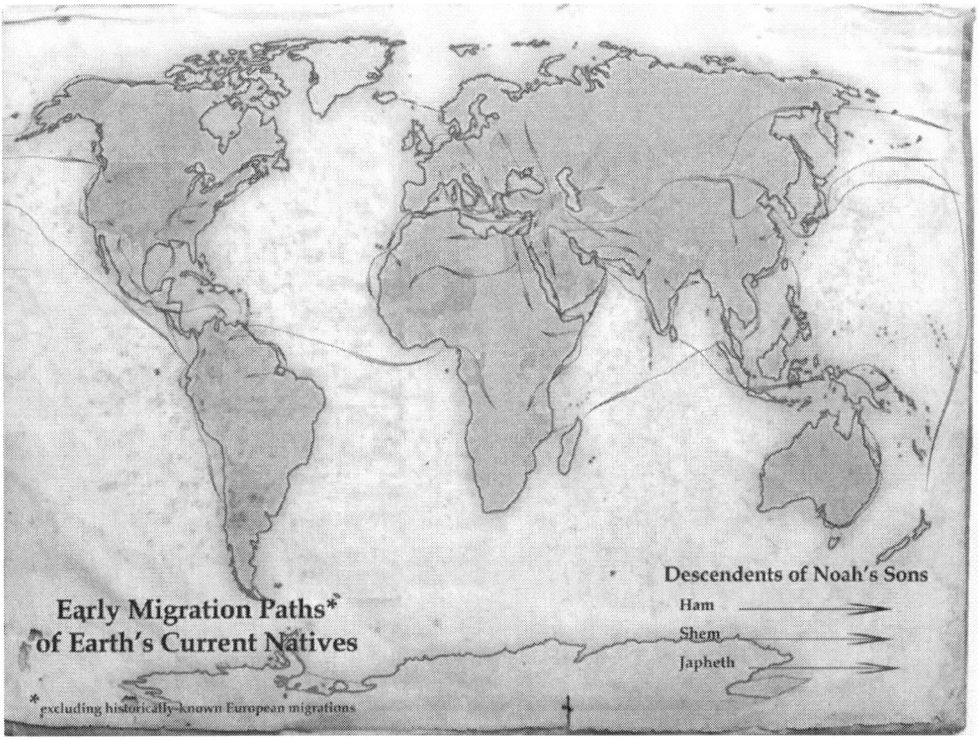

hieroglyphic scripts used to write the earliest texts; spoked-wheel wagons; agriculture and animal husbandry; side lugs for tying and carrying pottery; giant urns for storing grain; gray, sand-tempered pottery, some with eggshell thin styles that appear to have developed from metallic predecessors and chevron, swastika, and other patterns created by stamping and incisions; cremations, and urn burials. It is also characterized by the practice of animal sacrifice, recognizing among other gods a great Spirit (i.e. invisible God) who lives in the heavens along with memories of the great Flood. All this concerns the archaeology of those who survived the Flood.

Civilizations ended and buried by topsoil swept over them by the Flood

The archaeological sites of those who lived in the earlier civilizations who perished in the Flood have been mistakenly interpreted as elaborate religious burials. They are ordinarily found under mounds of soil as the case of the so-called Indian mounds of America, the barrows of Great Britain, the kurgans of Russia, and similar tumuli found throughout the world. This is because these sites contain elaborate burials as we might expect from houses and cities drowned by the Flood. Most of these dwellings are covered by mounds of topsoil, sand, and gravel that were eroded by the Flood then settled on these sites as the Flood subsided. We can conclude these are not ordinary burials but the ordinary dwellings of those buried by the Flood by a number of observations:

1. There are almost no places associated with these burials where the people buried there might have lived. Some explain this lack of occupational sites by supposing they belonged to seasonal-visiting pastoralists such as shepherds or herders. But most of these burials are in the floors of houses owned either by farmers or city dwellers. The truth is these burials *were* the occupational sites of those buried in them. This includes those buried in the trenches around Stonehenge, the trenches being the long-decayed ruins of wooden shelters. The *mastabas* beside the Great Pyramids in Egypt were however the mudbrick ground-floors of the wooden dwellings of those who built the Pyramids. The Pyramids, likely built for protecting treasures in the era before the Flood, were later remodeled as tombs for kings.

2. Ordinary sites where people once lived have worn out or broken artifacts, the types of things that are ordinarily tossed into refuse. But these burial sites have unworn and even unbroken artifacts as would have belonged to occupational sites that were suddenly ended. These artifacts also include various industries with product stores in both complete and incomplete states. Many of these are explained as 'votive offerings.' Anything can be imagined as having been a "votive offering." There is no puzzling artifact that cannot be explained by some or another religious interpretation.

3. Some of these graves are filled with great treasures of silver and gold. I got my first insight into the nature of these graves by reading Sir Leonard Woolley's account of the suttee funeral by which he explained why so many dancers and soldiers were buried at the same time with the King and Queen in the 'Royal Tombs of Ur.' Woolley complained of all the looting that must have taken place before he got his opportunity. No one buries such great treasures in unguarded tombs unless there be no one to witness the burial. I immediately recognized a party going on in the ancient city of Ur which Noah's Flood brought to a sudden end.

Woolley's workers were already standing in the ruins of the Flood when they dug through a 8-feet layer of water-laid mud which in 1929 Woolley announced to the world as evidence of the "Flood," then assumed but a local one. This stream of heavy mud would have been the first and lowest layer of silt to flow into the city from erosion caused by the waters that covered the Zagros mountains. Mud and topsoil covered the entire royal palace. Future settlers at the site of Ur would find this place more suitable for a garbage dump. The covered ruins of Ur became but one of the numerous tels throughout the Near East, most of them initially created by Noah's Flood.

4. Ordinary burials are mostly graves of the elderly. But, as the case of the young dancers in Ur, these burials contain people of all ages and of apparent excellent health. Numerous infants are found buried in jars which may have served as their beds or wash tubs.

5. Every place across the prehistoric world, people are buried with weapons and tools in caves and fields that look like ordinary campsites, in tombs that look like ordinary houses, and in necropolises that look like ordinary cities. Some lie on matting and platforms that look like ordinary beds, accompanied by household artifacts, many with personal weapons and jewelry.

6. Those in ordinary burials are placed in formally arranged positions. Not so those in these vast prehistoric burials, who are found in many casual positions as if by chance arrangement. Many of those buried in these sites are found in flexed positions, which is evidence of death by drowning. This happens due to arms and legs floating in water above the heavier torso. The corpses then settled in such flexed positions. Houses of great stones and timbers as well as cave dwellings kept bodies from floating away. But Otzi the Iceman, found in the nineties buried under a glacier in the Alps, was restrained by the strap that fastened him to his quiver.

7. Vast numbers of lowland burials from prehistoric sites are found under mounds or tels. Archaeologists tell us that Early Bronze Age remains in river valleys and deltas are too deeply buried for easy excavation. By contrast, Early Bronze Age sites found in highlands are either slightly buried or else found entirely above ground. This

is just what we should expect if the waters of the Flood eroded topsoil, sand, rocks, and clay from the mountains and highlands, sweeping this mixture of soils and rocks into the foothills and river valleys covering farms, villages, and cities. Many of these sites lie off the coasts of seas and oceans which contracted due to the expansion of the oceans resulting from water from the Flood.

8. The last thing I mention here as evidence of the world before the Flood comes from the light of our sacred texts, that is, it being a time of decadence and great violence. Archaeologists interpret the tiny vessels from this time as being used for drugs. There is also plentiful evidence of the use of great amounts of wine and beer. Excessive violence explains the numerous weapons, injuries to skeletons, and the thick walls of these Bronze Age cities. To judge the level of violence, one need only search the Internet for 'violence, Early Bronze Age' or 'violence, Neolithic.' If you do this search, you will be shocked by the level of violence, the chief reason that God brought this bloody world to an end.[8]

---

[8] c.f. Genesis 6:13

Chapter 3

HOW PREHISTORY GOT EXPANDED FROM A FEW THOUSAND TO HUNDREDS OF THOUSANDS OF YEARS

There does appear a problem with understanding these prehistoric sites as having been ended by the Flood. According to the present prehistoric chronology such sites as I describe as Flood burials extend for hundreds of thousands of years, from the Middle Paleolithic to the Early Bronze Age! It is however noteworthy that prehistoric eras are defined by archaeological type styles rather than successive strata at continuously occupied sites. This explains why the current prehistory of man jumps from Africa to the cavemen of Europe during the Ice Ages, then back to the Ancient Near East at the end of the Ice Age where begin agriculture and cities, but still jumping from site to site for thousands of years. Contrary to the claims, "the world's oldest cities" weren't continuously occupied for 10,000 years. The strata of Jericho, supposed the world's oldest city, show destructions at various times including the fallen walls of Joshua's Middle Bronze Age Jericho and a Flood-caused end of the Early Bronze Age city. (see Kathleen Kenyon's stratigraphic drawing of ancient Jericho on page 100)

Scientific archaeology got its beginnings in the early nineteenth century by those studying prehistoric sites in Scandinavia. Some of these ruins contained artifacts of bronze and iron, but others had only artifacts of stone. Christian Jurgensen Thomsen (1788-1865) decided the way to date these ruins from earliest to later would be according to whether they contain artifacts of stone, bronze, or iron. Thus, came what we know as the Stone Age, Bronze Age, and Iron Age that precede the ruins of the first historically known peoples. The Scandinavian archaeologists knew that civilization had spread to their own land from the lands of the ancient Orient but had little difficulty fitting these stages well within the about six thousand years era of man as can be calculated from the Bible. Then something happened to vastly expand the age of man.

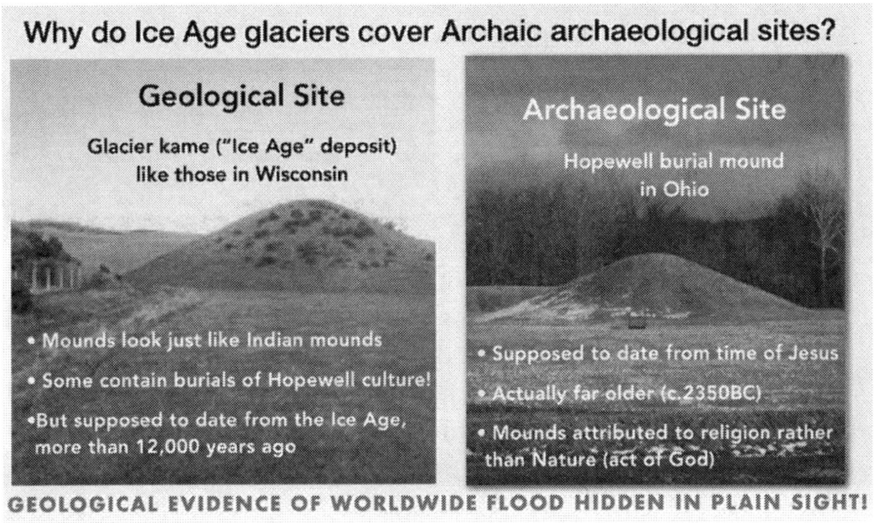

I mentioned in the *Archaeological Evidence* what clued me to the great problem in prehistoric chronology. It was the *Glacial Kame* prehistoric burials in America's upper Midwest which are essentially identical to the Hopewell burials along the Ohio River (see above).[9]

---

[9] Photos from Wikimedia Commons. Kame in left photo is in Scotland. Whether it also contains burials as do the kames in Wisconsin, see discussion p. 55

Archaeologists recognize the same culture is responsible for these burials, but the former are found under kames, supposed to have been deposited by glaciers during the Ice Age. The Hopewell burials which are part of the larger Adena culture are currently dated to around the time of Christ but the lowest burials in these mounds are of the much earlier Archaic cultures of America. The upper levels of these mounds which were used as burials by the historic American natives do date from the time of Christ though being of an entirely different material culture, one possessing bow and arrows and sand-tempered pottery. Because the historic American natives were accustomed to burring their dead in temple mounds, the top of these mounds would have been reasonable places to bury their dead if these mounds already existed, as had been the case since the time of the Flood.

But why would ancient Americans living in Wisconsin have gone to the great trouble of digging beneath Ice Age glacier deposits to build elaborate houses for burying their dead? And why would ancient Americans living in Ohio have spent vast amounts of time and resources to build elaborately furnished mausoleums only to cover them with dirt! The truth is, the mounds in both US states were created in the same way as the tels in the Ancient Near East, by eroded soil from the same great Flood. The Archaic cultures of America were ended by the Flood in the third millennium before Christ. So were Neolithic and Early Bronze Age cultures all over the world.

The Diluvium geological deposits before being re-envisioned as a long Ice Age.

We should not be surprised by my discovery because the geological evidence which we now associate with a supposed Ice Age was originally understood as evidence of a great deluge. As the name by which they were first described indicates, these *diluvium* deposits which we now attribute to the Ice Age were at the time of their discovery seen as evidence of a worldwide flood. Not all geologists agreed that the diluvium was the Noah's Flood of the Bible nor even that it covered every mountain but most agreed with Georges Cuvier, the world's leading geologist, that some great flood as vaguely remembered in the Flood accounts of many nations was responsible. *The American Journal of Science,* America's oldest and most prestigious scientific journal, was founded on diluvium geology.[10]

The stratigraphical understanding of the earth still being used in modern geology developed from seeing land arising from a receding ocean, an understanding originally based on the ninth verse of Genesis and taught by the German philosopher Gottfried Leibniz (1646-1716). The mining school known as *geognists* identified the first rocks to appear from the ocean beginning with granite as *Primary* strata. Rocks which had been compacted since the initial creation were called *Secondary* strata. The *Tertiary* strata

---

[10] For a review of the geological history behind this chapter, c.f. Martin J. S. Rudwick's two masterful volumes *Bursting the Limits of Time (2005)* and *Worlds Before Adam (2008)* published by the University of Chicago Press

were initially supposed to have contained the bones of animals who died in the Flood. Due to the discovery of the diluvium, the Flood was then understood as appearing in the topmost *Quaternary* washed up deposits which lacked the regular strata of the lower deposits.

The diluvium lay above the lower and very deeply stratified Primary and Secondary deposits including those which contained the fossils of fish and dinosaurs. They also lay just above the Tertiary deposits which contained the bones of early mammals. But they lay just below the *alluvial* strata created from the seasonal floods of present-day rivers. The diluvium occurs at various depths as might be expected from the deposits of a flood. Ordinarily, the diluvium is far deeper than the topmost alluvium, and unlike either the alluvium or the strata that lay below, the diluvium had no simple bedding pattern. What but a great Flood could explain such chaotic deposits and this sudden extinction of Europe's earlier wildlife.

The diluvium consisted of a variety of rocks and soils and contained the bones of extinct animals which aside from their larger size and geographical distribution were like those still roaming the earth. Among the types of soil found in the diluvium were gravel, sand, clay, the rich loess that cover the earth's fertile plains, and clay mixed with rocks called boulder clay. Today, the diluvium is everywhere covered by a layer of topsoil that formed after the diluvium was deposited some few thousand years ago. One can inspect the diluvium by digging through the topsoil that cover unimproved lands everywhere except at places like the eroded sides of mountains.

Though the source of the diluvium seemed to have been some flood, geologists did not agree concerning its cause. Jean DeLuc (1727-1817) attributed the flood that deposited the diluvium to an exchange (a "revolution") between earth and sea. D.G.D. Dolomieu (1750-1801) believed the diluvium was created when giant tidal waves either carried arctic birds south or swept tropical animals north. Among the animals which this Flood seemed to have made extinct in Europe were elephants, hippopotamuses, rhinoceros, hyenas, and apes. Horses and camels which then roamed in America also suddenly went extinct. Georges Cuvier (1769-1832), father of the science of paleontology, could only explain these extinctions as being due to a great flood which Cuvier also explained as a "revolution" of land and sea. Importantly, Cuvier's pointing to extinctions due to 3 earlier revolutions was key in establishing the geological acceptance of an ancient earth. Thus, it was the discovery of the true geological evidence of Noah's Flood that established geological acceptance of an ancient earth!

Inside the Kirkdale cave near Liverpool England, William Buckland (1784-1856) explored a hyena's den containing the gnawed bones of elephants and other animals now only found in the tropics. These remains were sandwiched between two layers of stalagmite, the upper layer covering the diluvium which drowned these cave dwellers. But even those like Jean DeLuc, Dolomieu, and William Buckland who identified the diluvium with Noah's Flood sought natural rather than supernatural causes.

Jean DeLuc pointed to the diluvium as evidence against James Hutton's theory of the earth as having always functioned in a uniform manner. As DeLuc explained, the present course of rivers cannot be so ancient. Else erosion would have long ago filled river lakes. The deltas where rivers flow into oceans being of limited extent were likewise evidence of the recency of the rivers. Contrary to Hutton's theory, steep V-shaped banks of rivers show that they could not be responsible for the wide U-shaped valleys through which these rivers meander.

Some of the most powerful evidence for a recent flood concerned two additional kinds of phenomena. One was the still-preserved frozen carcasses of now extinct animals found in the frozen tundra of Siberia. Some of these animals were eating at the time of their death. How could vegetation suitable for sustaining elephants be growing in such arctic conditions and how could their carcasses have not decayed aside from being somehow frozen immediately following their death? Had water from a worldwide Flood suddenly covered them, it would have turned into ice as the warmer waters receded.

The other evidence of a great flood in the recent history of the earth were rocks, some the size of houses, lying on the earth's surface which had gotten displaced as much as hundreds of miles from their geological origin. Ordinary erosion could not explain the displacement of these *erratics,* which didn't necessarily follow river valleys. But a great tsunami or flooding from the ocean might. Sir James Hall, a disciple of Hutton, suggested the *erratics* had once been under water where they had gotten frozen into icebergs. He believed that earthquakes in the middle of the ocean caused waves of translation to lift the icebergs from the ocean which then drifted onto land to the place where they currently lay. Thus, he explained how the rocks carried by these icebergs caused the parallel groves and scratches which he observed on the surface of the *erratics* and bedrock near Edinburg, Scotland.

Re-envisioning the diluvium as coming from the deposits of drifting icebergs

Charles Lyell (1797-1875) and his geological student Charles Darwin (1809-1882) became the chief proponents of the iceberg drift theory. According to them, *erratics* and various types of soil then attributed to a Flood came from the bottom of frozen oceans. But instead of being deposited on land by tidal waves, Lyell taught that the present lands and sea oscillate in relative elevation. As the oceans rise and the land sinks, great rafts of icebergs drift onto land where they dumped *erratics* and other kinds of soil that had gotten frozen into the icebergs. He therefore renamed the diluvium as the *drift.* According to Lyell, these great rocks were also responsible for shaping valleys and digging lakes, including the Great Lakes of America which lay just above the diluvium. It was a fantastic theory, but Lyell would bury these details in an encyclopedia of geological history and empirical observations that he had gathered from different parts of Europe and America.

According to Lyell, the slowly shifting elevation between the ocean and land was responsible for many floods over hundreds of thousands of years. In addition to the rising ocean floor being responsible for seashells buried in mountains, floods caused by rising oceans and sinking land lifted and carried great rafts of icebergs to the places where they dumped the *erratics* and what were then understood as diluvium deposits but which he saw as having gotten frozen into the icebergs. Thus, Lyell used the depth of the drift to estimate the gradual and uniform period required to deposit what was then understood as the diluvium, guessing it to be several hundred thousand years. These floods could not have been the same as Noah's Flood in the Bible, which however was precisely Lyell's aim. Lyell did allow that Noah's Flood might have been a local one limited to region of the Caspian Sea.

Lyell presented this theory in his three-volume *Principles of Geology,* the first volume published in 1830. Defending his drift theory against the diluvium, Lyell followed the Scottish minister John Fleming who proposed that the extinction of the large fauna found in the diluvium was due to human hunters. He pointed to George Scrope's failure to find evidence of the diluvium in the mountains of Auvergne but could explain steep and narrow valleys as entirely due to erosion. Scrope found volcanic tephra on the top of the mountains which Flood waters should have washed away. But diluvium lay deep below several flows of lava from Mt Etna in Sicily, indicating that it was deposited long before humans walked the earth.

These aspects of Lyell's *Principles of Geology* are overlooked in popular writings because the chief aim of this book was to defend James Hutton's claim that, given sufficient time, uplift of the earth stemming from intrusions from below together with ordinary erosion could explain all the earth's geological features. Lyell aimed to explained geology apart from extraordinary catastrophes as those claimed by *diluvialists* such as Cuvier and Buckland. He also challenged the claims of the *geognists* (see page 21) that the strata of the earth were created from gradually diminish oceans. But the chief obstacle to these claims was explaining away the most recent catastrophe, then believed responsible for the diluvium deposits.

Though ignored by critics and supporters alike, Lyell's use of percentages of living species to define the earth's recent geological eras (which Lyell himself invented!) wasn't based on mammals, the most recent and rapidly changing form of life, but on shellfish, the earth's oldest form of life which aside from still ongoing adaptations to new environments hardly changed over this recent period. The blame however should be shared with Georges Cuvier who used freshwater shellfish to define the entire range of geological periods based on fossils. In 1808, Cyprien Prosper Brard suggested that these freshwater mollusks were the result of a great diluvial quake that had mixed shells of various kinds. Thus, Cuvier's "last" Flood was sufficient to have mixed these freshwater mollusks with the more ancient fossils. But Cuvier needed his earlier "revolutions" to defend extinctions against Lamarck who supported Brard. Brard also pointed out that mollusks continuously adapt to changing salinity. Thus, they can be no reliable marker of time.

Lyell's new uniformitarian theory of the earth met with considerable skepticism among geologists, but not because it removed Noah's Flood from geology. The opposition was to his uniformitarian view which suggested that animals neither evolved nor went extinct. Lyell believed that dinosaurs were still to be found alive someplace in the still unexplored earth! More significantly, opposition to Lyell came from the fact that he denied that the fossil record showed evidence of increasingly complex life. Nor did geologists like using unpredictable volcanic eruptions as a reliable measure of time. Due to erosion, they knew that the diluvium wasn't to be found everyplace and especially on the sides of mountains. Moreover, how could the small populations of humans living before modern transportation have been responsible such great extinctions!

What geologists did like was how Lyell's uniformitarian science explained geology entirely by natural causes, the same as had attracted many to James Hutton's theory of the earth and would latter cause them to favor his disciple Darwin's theory of evolution. The diluvium would no longer be viewed as deposits of a single Flood but as an extended geological period currently known as the *Pleistocene.* Thus, Lyell used his drift theory to explain all the phenomena previously seen as evidence of a sudden and single worldwide Flood. Despite the rejection of his uniformitarian geology by most geologists and eventually by Lyell himself, Lyell saw his chief accomplishment as 'having rid geology of Moses.'

The Ice Age

In the meantime, a new theory emerged to explain the diluvium deposits that Lyell and Darwin were then attributing to iceberg drift. This theory had developed in Switzerland where existed both *erratic* stones and mountain glaciers. If the great displacement of the *erratics* had previously been attributed to the waters of a tidal wave caused flood, perhaps they could also be explained by a slow-moving frozen flood such as the glaciers which flow from high mountains like the one by which we are gathered. The popularization of this theory to explain the worldwide deposits of what was then known as the diluvium was the work of the world's leading expert on fossil fish. Louis Agassiz was already a well-known scientist when he became the pioneering advocate of the Ice Age.

Agassiz first presented his idea to the 1837 Swiss Society of Natural Sciences where it received a completely negative reception. The great Alexander Humboldt advised Agassiz to return to studying fossil fishes. Even so, Agassiz presented his idea in a memoir outlining the historical evidence for oscillating glaciers in Switzerland and proposed that vast ice sheets once covered Europe.

Agassiz shared the idea with his friend William Buckland, who by now was willing to abandon the diluvium theory for which he had been the chief advocate. What most impressed Buckland was that parallel grooves along bedrock could be better explained by glaciers sliding large *erratics* across bedrock than could Lyell's iceberg drift theory. But Aggassiz and Buckland assumed the same kind of Ice Age glaciers were also responsible for the diluvium that the latter had found in the tropics of southern Asia!

Buckland's support was key. By 1846, when Agassiz came to America to take a position at Harvard University, the Ice Age was all the rage in Europe and America. Agassiz soon embarked on a trip to Brazil to study the evidence of glacier deposits in the Amazon jungles, where he found them. As Buckland had found evidence of the diluvium in the jungles of Southeast Asia and in the highlands of the Himalayan Mountains, Agassiz concluded that glaciers had once covered the entire earth. Not surprisingly, Agassiz's teaching of what was then called *Snowball Earth* had the effect of cooling enthusiasm for an Ice Age. Thus, Lyell and Darwin continued to prefer their drift theory to such a non-uniformitarian theory as a recent great Ice Age.

In 1852, a scientific expedition to Greenland discovered that great ice sheets covered the island. Later in the century, ice sheets were discovered in Antarctica. As southern limits to northern ice sheets became accepted, so did the Ice Age. The diluvium does appear somewhat differently in the northern climates where less vegetation grows. Even to this day, the attention given to explaining an Ice Age now restricted to the colder northern regions deflects attention from the fact that the diluvium which by then was being called the *drift* is also to found in temperate and tropic regions.

Glacier scientists tell us that rocks freezing to the bottoms of glaciers operate like a great file, scraping the earth below them. Rocks do freeze to the bottom of glaciers, but the tensile strength of this part of the glacier is the greatest of any part of the glacier. Instead of scraping the ground, it will just retard the flow of the glacier. The geologist Sir Henry Howorth was appalled by the notion that ice sheets could defy the laws of gravity by flowing uphill, even over mountains, digging out deep lakes by dragging great boulders. It was however scarcely less difficult to believe than Lyell and Darwin's equally fantastic drift theory. That geologists could be convinced of an Ice Age was Howorth's 'Glacier Nightmare.' [11]

If not a dream, there was a feat of great imagination behind drawings of a miles-high wall of ice sliding towards what is now New York City. Though much lower, such ice cliffs as can be seen in Greenland or Antarctica are created when icebergs fall into the ocean, but where on flat earth might they fall? Glaciers do move towards the ocean through sloping valleys in the same way that unfrozen rivers flow toward the ocean. But to imagine a lateral force to propel icesheets across flat land, even uphill(!), glacier scientists must imagine snow somehow falling only in selected spots from which these icesheets are somehow supposed to spread. Even so, permanent ice covering builds up slowly and gradually in the frozen tundra and melts in the same slow way and without dragging rocks aside from what might flow from meltwaters. Agassiz's Ice Age theory however was not inspired by

---

[11] Sir Henry H. Howorth, *The Glacial Nightmare and the Flood* (Vol I, 1892; Vol II, 1893)

ice sheets covering flat lands, but by observations of *erratics* and deposits found in the valleys of mountain glaciers.

This summer, I had an opportunity to view just how these rocks tumble from this ice-capped mountain by which we are gathered. As the case of mountains throughout the world, Agri Dagh slowly grows causing rocks to fall from its higher levels. The winter snows cause the mountain's glaciers to grow and freeze the rocks beneath them. During summer melts, dams of water gather under the lower extensions of these glaciers. Our host Ahmet Ertugrul sent me a video that he recorded of a rapidly gushing stream of water shooting down this mountain from inside the glacier. The rushing stream of water was carrying rocks, large and small. It is gravity not glaciers that is responsible for most "Ice Age" deposits. Whatever deposited erratics along with loess, sand, and boulder clay across great plains, it could not have been glaciers. It could have been great amounts of swiftly moving water.

As to where to find the sources of this water one need only follow the sources of the *erratics.* In the northern and Alpine regions, they will be the source of the flow of what are now supposed as once Ice Age glaciers. But they can be found as well in temperate and tropic regions, resulting from the great underground quakes that led to the mixing of shells as pointed out in 1808 by Cyprien Prosper Brard (see p. 27 above). The emptying of the underground reservoirs that fed rivers and lakes prior to this event caused a considerable rise in the elevation of land which has been wrongly explained as being due to the melting of Ice Age glaciers. This also explains the rising of the mountains as the internal parts of the earth continue to empty their internal water, the same kind of virgin water that we now see expelled by volcanoes. It's why evidence of ancient vegetation that can grow only at low altitudes can now be found near the place where we are meeting, which is about one mile (1625 meters) above sea level.

Calculating the era since the Flood

Inspired by two of DeLuc's observations: (1) that river deltas date from the time of the Flood and (2) that silt lying above ancient monuments under river deltas might serve as a measure of time, Dolomieu traveled with Napoleon's troops to Egypt where he hoped to find a monument beneath the Nile Delta by which he might use to calculate the time since man first appeared in Egypt. Unfortunately, Dolomieu's work ended with the downfall of Napoleon, but others continued this work as published in the famous *Description de lEgypte* between 1809 and 1829. Based on the 72-foot depth of a monument found in the delta and his estimated mean rate of sediment build up in the delta of 2.5 inches per centuries, M. Rosiere calculated the time since the placement of the monument of to be about 30,000 years. That was at least 25,000 years too early, but no one in those days knew any better.

Other geologists believed they had a way of determining the time since the Flood deposited the diluvium, though it was coming to be seen as either the time since the drift or the Ice Age had made these deposits. As everyone acknowledged, the American Great Lakes rested on diluvium deposits. The erosion of Niagara Falls by the Saint Lawrence River had begun at Lake Ontario. Knowing the time since the Flood was a matter of estimating the rate of erosion of Niagara Falls and multiplying that by the distance of the Falls from Lake Ontario. In 1829, the British geologist Robert Bakewell used historical information from the people who lived near Niagara Falls to estimate the rate of erosion. Based on estimated erosion of 3 feet per year, Bakewell calculated that it had been about 10,000 years since Niagara Falls began eroding from Lake Ontario, then seen as being the time that had passed since the great deluge.

That was far too recent for Lyell, who had a need and thus a tendency to multiply the number of geological eras and expand the duration of each. Lyell himself created the many subdivisions of the Tertiary and Quaternary periods. To expand what Lyell called the most recent era, currently known as the Holocene, Lyell guessed the erosion to be only 1 foot per year to calculate at least 30,000 years. Lyell was no doubt acquainted with the calculations published in the famous *Description de lEgypte.*

This is important for understanding the *scientific* methods of the man most responsible for shifting the authority recognized by European Christians from the Bible to science and for wresting the sciences from the Christian universities to become secular professions. Though they appreciated his wide acquaintance with geological history and his reporting and addressing geological observations from all over Europe and America, even Lyell's close friend Oxford Professor Adam Sedgewick recognized that Lyell functioned more as an advocate with a case to win than as an unbiased observer of empirical observations, something admired in the earliest days of geology. Lyell was in fact a lawyer by training. In the nineteenth century, when geology served as the mother science from which most of the other historical sciences developed, Lyell functioned as did the politicians in the social circles of Europe where he sought audience. From the death of Georges Cuvier, another scientific politician who ruled the previous era, Lyell presided over geology and the sciences which sprang from geology. There is no need for putting *science* in quotes because government sponsored science, which is how the historical sciences are funded today, remains the most political of professions. Secular governments determine what is *scientific,* not so much according to truth but to whatever their laws will recognize.

In 1878, Alexander Winchell calculated the rate of erosion for St. Anthony's Falls near the beginning of the Mississippi River and found that the Ice Age must have ended only about 8,000 years ago. However appreciative that this famous European geologist took note of geology in America, American geologists such as Yale's James Dana didn't like Lyell so cavalierly dismissing their careful work. Strong opposition from American geologists who had a true empirical basis for their measurements restrained Lyell from expanding the Holocene, the time coming to be known as 'the end of the Ice Age,' to 30,000 years. I will explain below how Lyell agreed to compromise the time since the end of the Ice Age as being 12,000 years ago.

Chapter 4

DISCOVERING HISTORY, SCIENCE, AND THE ANTIQUITY OF NATIONS

Claims that humans have existed long before about six thousand years ago as can be calculated from the Bible for the creation of Adam date from before the Christian era. One of the most influential was that of the Hellenistic Egyptian historian Manetho who claimed an antiquity for the Egyptians as early as 35,000 years ago. Jewish and Christian chronologists including Josephus and Eusebius from whom we learn of Manetho's claims also rejected them. But the chief rival to the Bible throughout the Christian era was Aristotle's eternalist teachings. Following Aristotle, one could suppose that the evidence of man's infinite past had been erased and forgotten due to the ravages of time.

The biblical roots of history

In contrast to philosophy, which the Greeks preferred, historical understanding requires that people value, record, and preserve the memory of the people, times, and events of the past as we see in the case of the Bible. The Romanian historian of religion Mircea

Eliade (1907-1986) noted how the Western world's linear sense of history is grounded in the Bible. A distinction should however be made between, on the one hand the *historical*, and on the other hand, the temporal ways of understanding in the *historicism* of philosophical, theological, ideological, or scientific systems. The historical is contingent upon the acts of free agents whether these be God, other unseen powers, or men, as in the case of history known from the Bible and other chronicles. By contrast, historicist ways of temporal understanding are pre-determined by unvarying patterns or laws. For example, there is a determinist 'arrow of time' associated with the Neoplatonist philosophy of emergence that Augustine of Hippo employed in his famous *City of God.* The same is true of the Marxist dialectic of history. Likewise, the 'arrow of time' defined in the development of the earth as taught by the late Stephen J. Gould is presumed based either on causal laws or by unguided chance in the case of Darwin's theory of evolution.

All these are different from the contingent sense of history as we see in the case of the Bible. The latter is better described as an *arrow of history* that follows the world-changing miracles and events of the Bible: from the successive acts of Creation in Genesis to the events of the Flood and Exodus, through history-changing persons such as Abraham, Moses, David, Jesus, Mohammed, and Martin Luther. Due to these changes coming from actions of free individuals and thus not subject to discoverable laws, the latter can only be known through divine revelation or else from human testimony reliably preserved.

Absent the influence of the Bible, mankind reverts to some or another eternalist teaching, for example, a world of random eternal matter as taught by the ancient Epicureans, a world of eternal cycles as taught by the Stoics, or the world as having always existed as taught by Aristotle. As Eliade points out, most cultures from ancient times until now have lived in what everyone assumed an eternal or essentially unchanging world. What is currently referred to as the 'Western way of thinking' is a combination of the historical way of thinking rooted in the Bible and the non-historical *eternalism* that is rooted in Greek philosophy. These contrasting ways of thinking influenced Jewish, Christian, Islamic, as well as Western secular thought.

Until the rise of Islam in the eighth century, literacy and learning in the Latin speaking West were in the hands of the clergy, the people of the Book. Thus, in the West a *clerk* was the same as one who could read and write. But *science*, which began with Aristotle's teachings, was proscribed as sorcery during the early Christian centuries due to its associations with astrology, alchemy, and magic. Thus, the writings of Aristotle weren't translated into Latin, the language through which Christianity and literacy spread into Western Europe. Rather, from the time of Augustine in the late fourth century, an otherworldly Neoplatonism with its ascetic despising of the material world dominated Christian theology. It was therefore among the Arabs of the Islamic East where arose the beginnings of science, why we in the West use Arabic numerals. It was not until the Crusades that Aristotle's writings, which Nestorian Christians had translated into Arabic, were translated into Latin and Aristotle's science as developed by the Arabs became known in the faculties of the arts in Western universities.

In the thirteenth century, Thomas Aquinas shifted Augustine's Neoplatonist theology from a Platonic to an Aristotelian basis, a worldview which could now accommodate science. Even so, the

Thomistic worldview as depicted in Dante's *Inferno* lacked a true sense of history. Rather, Aquinas medieval worldview reflected the static feudal order deriving from pagan Rome. Thus, Renaissance artists depicted even the people of the Bible in Renaissance dress. Aquinas was however challenged by the voluntarist tradition in late medieval philosophy which, based on the greatness of the God of the Bible, rejected the idea that the human mind can understand the world apart from studying his individual creatures. This led to the empirical and analytical traditions that distinguishes modern from Aristotelian science.

The philological roots of modern historiography

It was the humanists of the Renaissance who recognized writings found in libraries as divided between *ancient, medieval,* and *modern.* By *ancient* they had in mind the classic civilizations of Greece and Rome. But with these major divisions began our modern world's historical understanding. A shift to the historical worldview of the Bible began in earnest with the back-to-the-Bible Reformation. Seeking sources to argue who better reflected the teachings and practices of early Christians, Protestant and Catholic historians together created a historical understanding of the Christian centuries.

The Scientific Revolution began as careful observations of the skies led to challenges of Aristotle's earth-centered cosmology; a paradigm shift known as the Copernican Revolution. Due to the Council of Trent endorsing Aquinas, Protestants were more willing to favor science that challenged Aristotle. Beginning with Joseph Scaliger (1540-1609) but more famously remembered in the chronicle of the nations of Archbishop James Ussher (1581—1656), this revolution gave rebirth to the chronology of nations neglected since the time of the church fathers. Unlike the humanists whose ancient history began with Greece and Rome, Scaliger gave attention to the ancient nations mentioned in the Bible such as Babylon, Persia, and Egypt.

Contributing to understanding the history of the nations of the Bible were the teachings of the Jewish rabbis from whom Christians learned Hebrew to translate the Old Testament. Because there are no word-for-word meanings between languages, translators must translate meanings rather than words. Not able to decide among the many allegorical meanings, translators such as Martin Luther preferred to translate literal historical readings as they learned from these rabbis. But how should OT prophecy be interpreted? From the beginning of the church, the OT was Christian history with the prophecies pertaining to Israel understood as fulfilled in a spiritual way by Jesus and the church, the Israel of God into which Gentiles are grafted. But following the disaster of the second century Bar Kokhba revolt, especially those Jews living in the Roman Empire had avoided messianic prophecy which however revived after the success of Mohammed as the rabbis began to expect a conquering messiah who would fulfil these scriptures.

By reading these rabbis, the Reformers also learned their *sensus judaica* interpretation of OT prophecy. Reformers like Luther and Calvin recognized that the Christian prophetic understanding had to be retained. However, a new interpretation of Romans 11 supported by Theodore Beza (1519-1605) and included in the notes of the influential Geneva Bible allowed some to see these Old Testament prophecies as pertaining to an end-time Israel consisting only of Jews who had converted to Christianity. It was an interpretation more in line with that of the rabbis except, regardless the philo-Judaism of these Protestants, the rabbis recognized that conversion to Christianity must mean an end to Judaism as understood by the rabbis. Perhaps due to Jews having long lived in separate communities, these Protestants re-imagined a religion as a race.

The shift in understanding Judaism from being a covenant people to a race, which also meant reading the Old Testament as Jewish rather than Christian history, was as profound as it was gradual. Instead of identifying with the faithful Old Testament saints as Christians had done from the time of Jesus, some Christians began seeking their identity with the classical Greeks and Romans whose languages they knew or else with the European nations whose vernacular language they spoke. Many today seem unaware that Jews spoke and read the same languages as Christians, while Jews in the Roman Empire used the Greek Old Testament until the Islamic conquest when rabbis began acquiring the Masoretic Hebrew text from Babylon, adopting it for Jewish worship. But as philologists began identifying nations with languages, the Old Testament seemed ever more a Jewish book about the history of those Jews who had rejected Jesus of Nazareth as the promised Messiah, looking instead for a more worldly messiah like Mohammed or King David.

One whose Jewish family had converted to Christianity was Isaac LaPeyrere (1596-1676) who sought help from Calvinist Prince of Conde, then after his conversion to Catholicism, from the Pope for the restoration of Jews to Palestine. LaPeyrere taught that all Gentiles came from men who lived before Adam. He justified this by pointing to stone tools used by the earliest inhabitants of France, while the sons of Adam had tools of bronze and iron. If Gentiles came from men who lived before Adam, their antiquity could be far older than Adam. The idea that there were men before Adam from whom Cain obtained his wife wasn't original with LaPeyrre. What was truly radical about LaPeyrre was his teaching that the Bible concerned only Jewish history, that Adam himself was Jewish while the Flood mentioned in the Bible was limited to the land of the Jews. This explained why the Egyptians, Chinese, Hindus, and even nations in America had histories dating long before the Flood.

As to why the Scriptures didn't teach the same, LaPeyerre claimed it was because they had been corrupted, a view soon to be accepted by some of LaPeyrre's associates including Thomas Hobbes, Richard Simon, and Spinoza. Explaining the Bible by means of composition criticism was soon to be tolerated by Protestants who, as already the case with the Catholics, were more than happy to base authority on the teachings of the established churches instead of the literal words of the Bible as preferred by separatists and dissidents.

The roots of the Enlightenment in the Puritan understanding of prophecy

The idea of an enlightenment that should follow the Dark Ages when 'antichrist' ruled Christ's church came from the Puritans' understanding of biblical prophecy. According to Thomas Brightman (1562-1607), this would be a time when Christ's church would experience its latter-day glory as promised in the Scriptures. Since the time of Augustine, teaching a future millennium had been regarded as heresy. But students of Cambridge University's great polymath Joseph Mede (1586-1639) saw this as leading to the millennial reign of the saints mentioned in the twentieth chapter of the book of Revelation and a time foretold in the book of Daniel when knowledge would increase.[12] Such theologians as these would form the Royal Society, the first modern scientific organization which would soon be headed by the great Isaac Newton (1642-1726).

These great polymaths sought accurate dates for events in the history of ancient nations like Babylon, Persia, and Greece for prophecies mentioned in the books of Daniel by which to accurately calculate the prophecies predicting Christ's appearance in the first century and also for the beginning of the millennium mentioned in the book of Revelation. Such prophetic history was tied to Four Monarchies symbolized in the book of Daniel, traditionally understood as the empires of Assyria, Persia, Greece, and Rome. Isaac Newton sought astronomical references in ancient texts by which to test his newly discovered laws of motion. Looking beyond the biblical text, he employed a euhemeristic approach to the pagan writings to seek an even more ancient chronology of the nations.

Newton, who secretly denied that Jesus was the 'Supreme Creator,' agreed that Jews and Catholics had corrupted the Scriptures which however served as a means of 'teaching the vulgar who are incapable of a nice philosophical understanding.' Newton's position kept him from openly teaching his unorthodox unitarian faith which however spread through Europe and America when the ancients obtained their victory over the moderns in the ranks of the

---

[12] Daniel 12:4

freemasons. The Enlightenment however was intended only for an elite. Not even Voltaire wanted skeptical views taught to the masses less they endanger the morals necessary for society's preservation.

A scientific understanding of the heavens and the earth begin when Bishop Thomas Burnet (1635-1715) explained the earth as developing from a cooling star (as Descartes taught) through 7 stages, from a star to (1) Paradise, (2) its corruption due to Adam's sin, (3) causing Noah's Flood, (4) leading to the earth's present state, (5) it's future judgment by fire, (6) Paradise regained during the millennium reign of Christ, until (7) the earth again turned into a star. As the late Stephen J Gould pointed out, despite seeming to follow the biblical arrow of history, Burnet was teaching a cyclical view of time while reducing the great world-changing miracles in the Bible to ordinary (natural) processes.

Newton admired Burnet's explanation of biblical history without employing miracles. Newton wished to avoid using miracles as evidence of the Bible's truth, preferring instead the Bible's foretelling of history past and future. Though he believed the text of the Bible had been corrupted, Newton thought that more accurate dates for the coming of the millennium might be determined by studying world events, much as do evangelical Christians today who concern themselves with end time prophecy.

Recognizing the latent Deism in Burnet's theory, William Whiston (1667-1742), Newton's chosen successor for his chair of mathematics at Cambridge, presented a theory of the earth based on Newton's theories but with God using comets to create the earth and later to bring about the Flood. Though the frequent signs of the coming millennium as preached by Whiston and the Puritans of his day would soon be ridiculed by the satirist William Hogarth, these biblical roots were not only essential to the birth of these sciences, but for establishing the historical nature of all of today's sciences.[13]

The classic understanding of the Enlightenment is seen as beginning with the spreading of the ideas of Isaac Newton and John Locke (1632-1704), both of whom advocating knowledge based on empirical discoveries according to the judgement of 'reasonable' men. Locke accepted the miracles of the Bible as having been based on reliable ancient witnesses but rejected Catholic claims of recent miracles. Even so, Locke shifted the basis of authority from the Bible to empirical reason from which are discovered the laws of Nature. Instead of basing morality on the authority of the Bible, Locke's disciples Shaftsbury and Francis Hutchinson taught that morality should instead be grounded on moral law. Beginning with these latitudinarians, enlightened Christians gradually replaced the authority of the Bible with natural theology and the science of morals.

The miracles reported in the Bible were supposed as but human astonishment over things not yet understood according to philosophical principles. Conyers Middleton (1683-1750) went so far as to claim that miracles had ceased after the time of the Apostles, initiating a controversy which exceeded that of his contemporary David Hume who denied even the miracles and prophecies of the Bible. The scandal was due to the Anglican Church resting its authority on the church fathers who reported miracles as

---

[13] James E. Force, *William Whiston: Honest Newtonian* (1985, Cambridge University Press) c.f. Forward by Richard Popkin

continuing in their day. But even orthodox Anglicans were scandalized by *Enthusiasm,* a reference to supernatural gifts such as prophecy being claimed by radicals such as Quakers and the French Prophets who appeared in England during the seventeenth century or whoever might claim to experience God.

How the Enlightenment embraced the classical pagans' geographical racism.

Shifting from the historical view of the Bible, the men of the Enlightenment not only reverted to the eternalist understanding of the world but also embraced the racist explanation of nations held by the classical pagans. The Greeks had attributed their superior race to the climate and geography of their Mediterranean lands, the barbarians living in lands too cold, too hot, too dry, or too wet, and thus suitable only for producing slaves.[14] Montesquieu (1689-1755) likewise saw those living in the southern countries as too hot tempered and those in northern Europe as too stiff and cold, but the land and climate of France was just right to produce temperate men like himself. Not only would the secular anthropology become based on racism, so would the new sciences of geography and climate.

Classic civilization however was also slave civilization. Due to Christianity, slavery (though not serfdom) had largely disappeared from Western Europe, but enlightened Europeans saw slavery as

---

[14] c.f. Benjamin Isaac, *The Invention of Racism in Classical Antiquity* (2013, Princeton University Press)

necessary when seeking wealth by establishing New World colonies. The men of the Enlightenment like David Hume believed that society needed to be led by virtuous men (i.e. those who defend their honor) who understood the necessity of this 'unfortunate institution.'

Christians understood every nation as having come from Noah's family. But the Enlightenment replaced Christian/Heathen (i.e. Gentile) nations with white/colored nations. The Enlightenment introduced racism in both monogenetic and polygenetic forms.[15] Monogenetic racism supposed a degeneration of the various races as they migrated to inferior lands. Polygenesis refers to separate origins of the races whether as separately created in the races' native lands or due to their inferior position along the hierarchical Great Chain of Being that informed the Enlightenment's eternalist understanding. As did the Greeks, the Enlightenment thinkers equated the races' capability according to their geography.[16]

These men knew that great civilizations existed in the Orient when Europeans were but illiterate barbarians, but they regarded the Oriental races as having long ago reached their pinnacle of development. They attributed what some believed as the inability of emancipated Jews to adopt German culture as due to the inferior desert lands that produced the Semitic languages.

---

[15] Emanuel Chukwudi, editor, *Race and the Enlightenment, A Reader* (1997, Blackwell Publishing)

[16] c. f. Sara Eigen, Mark Larrimore, editors, *The German Invention of Race (2006, State University of New York Press)*

Due to the men of the Enlightenment having adopted the worldview of a temporally unchanging world as held by the classical pagans, geography replaced the past for understanding the Bible. Thus, to travel to the lands of the Bible was to travel to the world of Abraham and the people of the Bible. To learn more about the Bible, the Gottingen scholar J.D. Michaelis (1717-1791) convinced the King of Denmark to sponsor an expedition to Arabia conducted by a team which included the lone survivor of the expedition Carsten Niebuhr (1733-1815). Some may wonder why Yemen was the ultimate destination of Niebuhr's travels and why the German composition critics used the sheiks living there as the model for Abraham. The reason is simply that Michaelis, however wrongly, believed that the Hebrew language had derived from Arabic. He also believed that Yemen was the place where the Arab language was preserved in its pure original state.

Missionaries were forbidden to join the expedition less the peoples of the lands where they traveled become too familiar. Nor was Michaelis interested in the Islamic faith. Instead, members of this expedition were instructed to maintain an unattached objectivity when peering into the inscrutable minds of the Orient as if they were objects for scientific study rather than people like themselves for Europeans to get to know. By 'Orient,' they had chiefly in mind the lands of the Bible, but the 'inscrutable Oriental' would also be extended to the people of Far East Asia. Immanuel Kant taught that the very minds of the different races were biologically determined according to the geography of races' homelands. Accordingly, Michaelis engineered a deliberate attempt to make the customs, people, flora, and fauna from the lands of the Bible strange, also to make the Bible strange even to pious Christians, something that could only be understood by scholars like Michaelis and other "higher critics" of the Bible. [17]

The Enlightenment's esteem of James Ussher's chronology of the nations

Due to the classical Greeks' positive writings about the nation, Egypt was an exception to the Enlightenment's low view of the lands of the Bible. It was also due to another set of writings that challenged the history of man as taught in the Bible, the *Hermes Trismegistus* (thrice great Hermes), then supposed knowledge possessed by Egyptians living before the time of Moses. These *Hermetica*, which Ficino translated into Latin during the Renaissance, seduced most of the greatest Western minds including Isaac Newton. Newton taught that Noah passed the history found in the earliest chapters of Genesis to his grandson Egypt from whom this ancient knowledge was passed to Moses. Newton believed that writing was not invented until the reigns of David and Solomon. Newton saw his famous principles of natural philosophy as a restoration of ancient wisdom, a view that divided the men of the Enlightenment between ancients and moderns.

The great erudite Joseph Scaliger had already proven that the *Hermetica* were but Gnostic writings dating from the early Christian centuries. Scaliger also challenged Manetheo's Egyptian chronology. But to the dismay of Christians, the date that Scaliger

---

[17] c.f. Jonathan Sheehan, *The Enlightenment Bible* (2005, Princeton University Press), chapter 7

supposed for the beginning of Egyptian history was older than the creation of Adam according to the Hebrew text used to translate Christian Bibles. Thus, Isaac Newton who used chronology and ancient astronomical observations to test his new theories supported the chronology of Archbishop James Ussher. Due to Newton's endorsement, Ussher's chronology of the nations carried authority throughout the Enlightenment even with Deists.

Others supported Aristotle's teachings that mankind had always existed. The anonymous book *Telliamed* written by Benoit de Maillet, France's ambassador to the Ottoman Empire, and published in 1748 made this very claim. Following Descartes in supposing that the earth was originally covered by water, de Maillet explained the continents as being slowly formed from a gradually diminishing ocean over a period of at least two billion years. De Maillet believed that the earth was eternal and thus never created, but as also the case with Aristotle, he believed that man himself in some form or another had always existed. De Maillet taught that man originally existed as a fish in the ocean but had evolved to his current state.

Because of widespread acceptance of Archbishop Ussher's 4004BC calculation for the beginning of Creation even by anti-Christian Deists such as Voltaire, some have wondered how orthodox Christians like DeLuc, Dolomieu, and Buckland could so easily champion the great antiquity of the earth. This is because we look at this in an anachronistic, post-Darwinian light. Their battle was against Aristotelians who taught that the earth including mankind had always existed and more immediately against Deists like James Hutton who denied that the earth showed evidence of its creation or of any other miracle. As a Deist, Hutton could just as easily as

Voltaire have accepted a 4004BC date of the earth being created instantly and fully functioning much as do today's young earth Creationists.

By contrast, it was the Progressive Creationists' discovery of an ancient earth prior to man that established the fact of the earth's Creation including specific acts of Creation mentioned in the first chapter of Genesis. Cambridge University's Adam Sedgewick, a Progressive Creationist who defended diluvium science, discovered the Cambrian Explosion, the sudden and miraculous appearance of fish in the seas. This was followed in the fossil record by evidence of the animals who live on the land, much like the sequence in the Bible. Unlike the eternalist views of the pagan Greeks, the Bible mentions the earth as having existed prior to man and man as being the very last thing created. Geology seemed to be gradually confirming rather than contradicting the Bible while the light of the Bible was essential to the creation of the historical sciences.

Calculating the beginnings of Egyptian civilization

Aside from pious Christians, nineteenth century scientists believed that Egypt was the oldest civilization. Whether Egypt was under water due to the shifting elevations of land and sea as Lyell believed, or from melting waters of the Ice Age, geologists concluded that the flood waters must have ended before civilization began in Egypt. Taking his cue from Dolomieu's work in Egypt, a solution to the beginning of Egyptian civilization was worked out in the early eighteen fifties by Lyell's father-in-law, Leonard Horner, then President of the London Geological Society. From London, Horner directed drillings into the Nile River delta to determine the time that had passed since the erection of Egypt's oldest monuments. The time would be calculated from the silt that had built up in the Nile delta covering these monuments. Based on the depth in which burnt bricks had been found beneath the Nile Delta, Horner concluded that it had been 13,371 years.[18]

Now, that's remarkably close to the beginnings of Egyptian civilization which the Greek historian Herodotus calculated by what he learned from an Egyptian priest.[19] The Egyptian priest told Herodotus that the Egyptians had enjoyed a continuous priesthood of 341 generations from the time of Egypt's beginnings as represented by the wooden statues that each high priest had carved for himself. Calculating 3 generations for each 100 years, Herodotus determined that Egyptian civilization had begun 11,340 years ago. Around 1850 when these soundings were conducted it had been more than 2,000 years since Herodotus learned this, bringing us very near to what Horner calculated.

Herodotus's calculations based on Egyptian myth were however hardly more fanciful than Lyell's *scientific* measurements as the latter multiplied a 6-inch per century build up to a depth of 60 feet where Horner had found burnt brick(!), arriving at a neat 12,000 years for the beginning of Egyptian civilization. This span of time continues to define the Holocene, the last and most recent

---

[18] An excellent history of Horner's work has recently been published. c. f. Meira Gold, 'Ancient History and the geological antiquity of man, 1847-1863,' *History of Science* 2019, Vol 57(2) 194-230

[19] *Herodotus, Book Two,* p 186 of 1954 Penguin Classics translation

geological era.[20] Should anyone doubt that scientific dating continues to be calibrated according to expectations, know that more than 70 years since the beginning of radiocarbon dating, an era of 11,650 bp for the Holocene is recognized by the International Commission on Stratigraphy!

---

[20] Charles Lyell, *The Geological Evidences for the Antiquity of Man* (2004 Dover edition), p. 30

Chapter 5

Discovering the Antiquity of Man

Charles Lyell long remained a strong proponent of the belief then common among leading geologists that the diluvium, which he now called *drift* but which others were interpreting as evidence of an Ice Age, included no evidence of man. Even those such as Cuvier and Buckland agreed that at the time of the diluvium Flood, mankind had probably not spread outside of the lands of the Bible. Cuvier had examined several claims for the fossils of men who may have died in the Flood. The most famous of these, Cuvier determined to have been the remains of a giant salamander. Another possibility was the skeleton of a man found in the Caribbean which was encased in limestone. Due to finding organic remains in the skeleton, Cuvier determined it not to have been an actual fossil. Functioning as the virtual pope of natural history, Cuvier turned from initial open-mindedness to dogmatic insistence that there are no human fossils.

Buckland had himself found a red-painted human skeleton among the extinct fauna in the caves of Paviland which he himself had excavated. Accepting Cuvier's dictum concerning there being no human fossils, Buckland determined his 'red lady of Paviland' to have been a later burial intruding into the layer where he found extinct fossils. There were other reports of human artifacts or remains among the extinct animals within the drift. Lyell was particularly concerned to dismiss them less the scriptural geologists who he so despised claim them as victims of the Flood.

Finding men in the drift with extinct animals

Among those which Lyell had dismissed were chipped flints found in drift deposits near Abbeville, France that had been carefully excavated by Jacques Boucher de Perthes. Jacques de Perthes exhibited these in 1838, calling them *diluvium* remains. Because of the cold reception that Lyell and others had given his *diluvium* remains, he published them again in 1848, this time calling them *antediluvian* remains. This slight change in terminology allowed that these human artifacts no longer need have been from the Flood but possibly from far more ancient times if the diluvium were the deposits of hundreds of thousands of years as Lyell himself believed.

In 1858, an English geologist named Falconer was visiting the area and saw the chipped flints. After arriving back in England, he shared his observations with Sir John Prestwich and John Evans, recommending that the two of them visit the site in Abbeville, France. Prestwich and Evans did just that, spending much time at the site and confirming that without doubt there was evidence of man in the same drift deposits as had been found extinct animals. Evans wrote to his sister, 'the flat axes and implements found among the beds of gravel were evidently deposited at the same time with them in fact the race of men who existed when the deluge, or whatever was the origin of these gravel beds took place.'

As his letter to his sister reveals, Evans understood that these might be evidence of men who perished in the Flood. But Lyell was managing the marketing of these discoveries. On May 26, 1859, Prestwich read his paper to the Royal Society entitled *On the Occurrence of Flint Instruments with the Remains of Animals of Extinct Species in beds of a Late Geological Period at Amiens and Abbeville and in England at Hoxne.* Prestwich's lecture was attended by Lyell, Huxley, and Faraday. Shortly afterwards at the British Association of Science, Lyell giving the Presidential Address announced that he was 'fully prepared to corroborate the conclusions ... recently laid before the Royal Society by Mr. Prestwich.'

This would be the same year that Lyell's disciple Charles Darwin would publish his theory of evolution. In his first book, Darwin avoided treating the evolution of man, but in 1863, Lyell published his influential G*eological Evidence of the Antiquity of Man,* which prepared the way for Darwin's 1871 *Descent of Man.* What should have been heralded as the first evidence of men who perished in Noah's Flood has instead long been heralded as the first evidence of the antiquity and evolution of man!

How geologists missed the fact that the deposits from the Ice Age (formerly, the diluvium) cover prehistoric burials throughout the world

It is impossible to overstate the profound consequences for our world due to neither geologists nor archaeologists recognizing that the diluvium ("Ice Age" deposits) cover prehistoric burials throughout the world (see graphic with mounds, page 19) and due to *anthropologists* distinguishing those whose artifacts were found in the drift from those buried under archaeological mounds created from the same kind of deposits as those of the drift. I am reluctant to call those who did this *archaeologists* because they based their judgements almost entirely on studies of living *savages,* to use their terminology. These men were unabashedly racist as were most anthropologists and ethnologists of their day who weren't Christian missionaries taking the gospel to all the nations who descended from Adam and Noah.

There was too little discussion as to whether those few human bones and artifacts found inside caves and along the banks of rivers may have been contemporaneous with or even belonged to those who built the megaliths and dwellings found under tumuli on every inhabited continent. The failure to do so led to belief that mankind was far older than could be harmonized with the Bible.

There were however several things conspiring to cause this important matter to be so neglected. The most important reason for judging the remains of man found in the diluvium as being of great antiquity was of course because Charles Lyell had stretched the era of the diluvium deposits from a single year to a Pleistocene era of hundreds of thousands of years, which would eventually turn into an extended Ice Age of several million years. Thus, anything found in the diluvium, which Lyell was then calling the *drift,* could be presumed to be of great antiquity.

Until this present author, it seems that no one has ever noticed that "drift" also covered archaeological mounds which no one presumed to be of such great antiquity. Jean DeLuc had pointed to the layer of peat on top of these mounds as evidence for the time that had passed since the earth's "last revolution" which he taught as the biblical Flood. He noted that it was the same depth as the peat that covered all the land, which included diluvium deposits lying just beneath a layer of topsoil. Because DeLuc and the influential Cuvier believed that the antediluvian world now lay under the oceans must have caused some geologists not to think of those earthen mounds as having been produced by the Flood. This view of the Flood as having been due to a revolution between land and ocean also caused geologists to expect to find saltwater seashells in the diluvium, which they didn't find. Rather, aside from some coastal areas there were only freshwater shells as would however be the case if the waters from the Flood came from springs beneath the land as taught in the Bible and as proposed in 1808 by Cyprien Prosper Brard due to his paleontological observations (see page 27).

Still another important reason for not connecting the men found in the drift to those buried under the mounds was the dualistic idealism of Deistic beliefs that Lyell shared with leading nineteenth-century scientists. Largely due to the success of Darwin's theory of evolution, physical materialism has ruled twentieth century science, but an idealistic dualism dominated nineteenth century science. These Deists, who were the leading European scientists, taught a radical separation between Nature and man, the latter being in so many ways an exception to Nature. Thus, they excluded from Nature everything associated with man including the earthen mounds that covered man's earliest remains.

In any case, George Cuvier had assured them that humans didn't exist in Europe during the era of now extinct animals. Cuvier also noted how the pictures, sculptures, and embalmed animals that Napoleon's troops had brought back from Egypt resembled current species. According to Cuvier, animals had not changed over the long period that had passed since the time of these ancient Egyptians. While aiming to refute Lamarck's theory of evolution, this might also suggest that no new species of animals had appeared and perhaps that only local extinctions had occurred since the era of mankind. Recall (page 27) that Lyell believed that dinosaurs might yet be found in the still to be explored parts of the earth.

Cuvier notwithstanding, there aren't lions in the lands of the Bible as in the time of King David. Animals continue to go extinct while new species appear due to changing environments. There was a radical change in the climate and land following the Flood to which the bodies of animals would have quickly adapted as they spread across the face of the earth. It wasn't the Bible but their idealistic Deism that caused nineteenth century scientists not to wish to see extinct animals during the era of man.

When Deist geologists such as Lyell did acknowledge the remains of man as being in the drift, instead of seeing these as part of human history, they preferred seeing these cavemen as part of Nature where they found these remains. *Nature,* the object of their study, was exclusively the great outdoors: the mountains, the oceans, and the great wilderness. Scientists such as geologists and biologists studied Nature, which Lyell reckoned as being fixed and having no history. Archaeology on the other hand concerned the study of man. It developed from the humanities and from the study of books, languages, history, art, and antiquities which would include those stone tools found under the mounds. To protect the dignity of man, Lyell aimed to separate man and his history from Nature. Whereas the mounds belonged to the ancestors of civilized Europeans, anything found in the drift was part of Nature.

The chief reason for the rise of Deism during the Enlightenment was to exclude miracles, which caused Deists to reject a literal reading of the Bible and was something that composition critics aimed to explain away. Even so, it is important to understand that Deists needed their monotheistic Creator to explain the order and unity of Nature as would be required for science. They supposed however they didn't need and certainly did not want the interference of some authority whether the Catholic Church or, as some Protestants insisted, the literal words of the Bible. At least until Darwin put forth his theory of natural selection in which case order is presumed to be a mere appearance, science needed the presumption of a monotheistic Creator to expect that the Creation could be understood according to science and reason.

Awkward as it may be to the current debate, Deists also loved and needed *Intelligent Design* to guarantee the success of reason and science. When pondering the beauty of Creation, Deists such as Voltaire were piously devout concerning the Creator. They also knew this was included in the teachings of the Bible as were rewards and punishment in the afterlife, which they believed much as do liberal Christians today. However, as in the case of Thomas Jefferson, they regarded the miracles of the Bible as due to ancient superstition. Though they rejected literal belief in the Bible, not so different from today's prosperity teachers, they tended to ignore the reality of suffering and the presence of natural and spiritual evil.

Though not himself a Deist, the idealization of the natural Creation as famously and dangerously taught by William Paley lies behind Charles Darwin's argument for evolution. It caused Darwin not to expect pain and suffering as part of the ideal world which a wise and good God would have created. As also the case of the God of today's prosperity teachers, the Deist God had created Nature for man's life and enjoyment. The ideal Nature of the Deists was of course an imaginary world which differs from the kind of world as taught in the Bible and as we observe.

The ideal Created world of the Deists ignored the curses on mankind and Creation which the Apostle Paul taught as being due to the sin of Adam. Moreover, the elimination of pain, suffering, evil, and injustice is a promise for the incorruptible New Creation. For the present age, trouble, injustices, and suffering are for the discipline and maturity of those born into the corruptible first Creation. Though the Lord God found had found his Creation good for his intended purposes, only in the protected Garden of Eden where the Lord met with man and where the Lord had planted the Tree of Life did there exist rest and a way of avoiding death. But instead of God aiming to restore the Old Creation, the scriptures tell us that the heavens and earth in their present form will pass away.[21] Early Christians saw the Ark of Noah as an allegory for how the church would survive the world's destruction on the great and terrible Day of the Lord that would accompany the Second Coming of Jesus.

According to Darwin, God would not have created a world that included cruel animals and suffering, or the untimely death of children like his daughter Anne. In Darwin's opinion, God would not have designed Nature red in tooth and claw as Darwin determined to be the case. The *Origin of Species* is as much a theological as a scientific work. Darwin's ideal creation is however much like the

---

[21] Psalm 102: 25-26; Isaiah 34:4; Hebrews 1:10-12; Matt 5:18, 24:35; 2 Peter 3:10; Revelation 6:14

view of the pre-Fallen world as taught by today's young earth Creationists!

Finally, but not least important for the "discovery" of the antiquity of man was the signal influence of Lyell and Darwin. This was not only their methodology of supposing that ordinary erosion could explain why evidence from supposed enormously long periods of time is completely lacking in the fossil record. It was their higher social status than was the case of most scientists. Trigger writes concerning the Canadian geologist John William Dawson, who was Principal of McGill University in Montreal from 1865 to 1893:

> Dawson accepted the association between human remains and extinct animals but argued these associations confirmed the recency of the Pleistocene gravels in which they were found. On a trip to Europe in 1865 he inspected the geological deposits of the Somme Valley and described his mentor Charles Lyell as taking 'very good-naturally' his opinion that evidence was lacking "of the excessive antiquity at that time attributed to [these formations] by some writers."... He also maintained that North American ethnographic evidence revealed that people who used the best-made stone implements also used the rudest and that the developmental sequence found in Europe might represent idiosyncratic local trends or the accidental interdigitation of neighboring contemporary groups with different cultures. From this he concluded that there was no evidence that cultures at different levels of complexity had not coexisted throughout human history. [22]

Trigger concludes, '...in the nineteenth century it was easier for his opponents to ignore his objections that refute them.' Lyell's uniformitarian science together with Darwin's theory of evolution

---

[22] Trigger, Bruce G., *A History of Archaeological Thought* (1989, Cambridge University Press), pp 102-3

by natural selection were carrying the antiquity of man in their wake.

The roots of evolutionary theories

As indicated by this 1838 drawing of *fossil man*[23] and by the anonymous and controversial *Vestiges of the Natural History of and Creation* published in 1844, not to mention the evolutionary theories of Lamarck and Herbert Spencer, belief in evolution long predated Darwin. As the drawing suggests, these wild men of Nature may have been as nearly animal as human. Whether these half-human creatures were still to be found, the theory of human evolution would require less than human creatures and an extended era of prehistoric man to have preceded the era of man's civilization.

Fossil Man (1838)
conceived as dark-skinned, simian, and furry

Arthur O. Lovejoy points out that the theory of evolution came about by temporalizing the Enlightenment's Great Chain of Being.[24] Carl Linnaeus had already assigned man to

---

[23] Pierre Boitard, "L'Homme Fossile" (Fossil Man) in *Magasin Universel (1838)*, republished in Martin J.S. Rudwick's *Earth's Deep History (2014, University of Chicago Press)*, p.205

[24] Arthur O. Lovejoy, *The Great Chain of Being* (1964, Harvard University Press)

the order of the Primates, essentially classifying man as an animal. Between these groups, he added Troglodytes, the wild men of legends. The Great Chain of Being had developed from Aquinas's Aristotelian version of Neoplatonism. Aquinas applied his order of the world to Heaven, Hell, and Christendom, but the Age of Exploration beginning in the Renaissance acquainted Europeans with previously unknown places, peoples, animals, and plants.

The new discoveries needed to be named and classified. Thus, began the process of collecting and classifying along the Great Chain of Being, or rather according to numerous versions of the Great Chain of Being which classified all manner of minerals, flora, and fauna. It was difficult to know whether a certain animal may have been a new species or merely a variation of an already known species. Trying to solve this problem, Johannes Buteo (1492-1572) began the binomial classification system used by Linnaeus by trying to determine the number and kinds of animals that could fit on Noah's Ark. The Latin terms *species* and *genus* were used interchangeably before genus became the more general term.

Even so, proponents of different versions of the Great Chain of Being remained divided between those who regarded species as real and those who supposed a continuity between all species as would lead to a theory of evolution. This problem of defining species besets evolutionists to this day because they refuse to distinguish adaptations within the same species from their claims of evolution. The former would be limited to species that can produce fertile offspring just as the Bible defines as the kinds of created species. Acknowledging the distinction would of course refute evolution, though one could still hold to an eternalist idealism as did Louis Agassiz and John Owens, who were the earliest opponents of Darwin's theory of evolution. Evolution however required the temporalization of species which required the light of the Bible, whether for a historical or historicist (see pages 35-36) view of the earth and the cosmos.

Lovejoy seems not to have understood just how the Great Chain of Being became temporalized or how the Bible was needed for that. The first step in this process was the temporalization of cosmology and geology. Not surprisingly, the beginnings of cosmogony and geo-history came from studying the earth in the light of the Bible. This began with Nicolas Steno (1636-1686) who laid down the principles of stratigraphy and showed how fossils were of formerly living things whose disposition Steno attributed to Noah's Flood. Steno believed that the earth was no older than about six thousand years. So did John Woodward (1665-1728) who explained the order of fossils as being sorted according to their specific gravity after God suspended Newton's principle of gravity to cause Noah's Flood.

William Whiston (1667-1752) who I mentioned as explaining the history of the earth by comets (see page 43) allowed that the Creation days may have been a full year in duration due to God not yet having given the earth its daily rotation. Whiston's extension of the duration of the days of Genesis influenced Comte Buffon (1707-1788) who explained the earth as developing from a cooling star. According to Buffon's measurements of the time needed for cooling ingots of iron, the six days of Genesis might have taken more than 75,000 years.

Understandably, by the end of the eighteenth century such cosmological speculations were becoming unpopular. Thus, Jean DeLuc insisted that *geology* (a word invented by Deluc) be based on actual observations of the earth. This meant that geologists needed to visit far off lands and mountains to obtain their firsthand information. By the nineteenth century, new ways of traveling made this easier, though it required leisure and money that favored well-connected and affluent men like Charles Lyell.

Lyell not only believed in using actual data, he insisted that geology be explained entirely by processes of the same kind and operating at the same rate as those of the present day. By imposing such *a priori* dogma, Lyell rejected *a posteriori* empiricism. As a Deist, he refused to recognize any extraordinary events in the earth's history as would be the case of a worldwide Flood no matter how much evidence may support them. However, one of the reasons for Lyell's conversion to uniformitarianism was to counter Lamarck's theory of evolution which Lyell saw as undermining the dignity of man. Because of his concern for human dignity together with his uniformitarian principles, Lyell rejected Darwin's theory of evolution by natural selection until his angry disciple pressed him. It was a difficult concession on the part of Lyell because Darwin had based his theory of evolution on the non-uniform geology of the Progressive Creationists whose geology Lyell had long opposed. However atheistic, Darwin's theory of evolution by natural section owes its progressive direction to the Bible!

The Bible was likewise necessary for the temporalization of mankind. The Enlightenment's secular view of man began with John Locke who envisioned society as beginning in an egalitarian state, a time when men lived close to Nature. According to Locke, men learned to govern themselves by a social contract such as the earliest English colonists in America had pioneered and as the Whig-led Parliament recently negotiated with the new English King. Locke opposed patriarchy and the divine right of kings. Though secular, his social contract reflected the narrative of the Bible, a time of innocence until man sinned. According to Locke's narrative, "sin" was the presumption of the divine right of kings and patriarchy as then existed in the Old World but something that the American colonists in the New World had found unnecessary for governing.

Likely, Locke's view of earliest man was based on the self-governing towns of praying Indians about which the missionary John Elliot had recently published in London in a tract entitled *No King But Jesus.* The life of most of the native Americans was however more accurately described by the atheist Thomas Hobbes against whom Locke was writing. Brutish and short, it was everything but egalitarian, the very reason these towns of praying Indians were so hated by the chiefs who tried to destroy them in King Phillip's War.

However secular his methodology, Locke's view was one of restoring society to an original state of innocence. Thus, economic and scientific progress as happened with the Industrial Revolution was necessary for a view of evolution. Adam Smith was closely associated with the Scottish Primitivists who were influenced by Locke, but who envisioned man in economic rather than political terms. According to Henry Homes, who became Lord Kames (1696-1782), primitive man developed in four stages, from (1) hunter-gather to (2) herdsman and (3) agriculturalist until (4) participating in a market economy. In Lord Kame's view, man's initial state was more like that of Cain who was condemned to wander, while the earlier state of Cain and his brother Abel respectfully as agriculturalist and herdsman followed instead of preceded this initial state. Thus, instead of recognizing the reality of degeneration due to the sin of Adam, this view led to progressive unilinear evolution

Lord Kames however taught a polygenetic view of the origins of man, as did his cousin David Hume. But Lord Monboddo, James Burnett (1714-1799), friend of Erasmus Darwin, taught a monogenetic understanding of man by extending evolution to human biology. However incredible it may seem to us, James Burnett taught that humans are born with a tale which midwives discretely remove before presenting babies to their mothers. Surely, Darwin conceived the origin of man from naked apes from what young Charles learned from his grandfather. This might also explain why Voltaire, who was familiar with these teachings believed that those living in deep Africa were the products of women raped by orangutans.

Immanuel Kant, on the other hand, was concerned about the degenerative effects of the interbreeding of races. According to Kant, blond-headed Germans like himself, the purists of the original race of white men and hence the most excellent, needed to keep the white race pure through proscribing miscegenation. As mentioned above (page 47), Kant believed that inferior non-German lands determined the very minds and souls of the races from those lands. Kant's racism was based on his view of the colored races as having degenerated from the white race.

The current meaning of *primitive* as being like a presumed low state of original (primitive) man derives from these Scottish Primitivists. It became a principle informing the evolutionary view of man, even if some of the Scottish Primitivists viewed the races as having separate positions according to their hierarchical Great Chain of

Being or else viewed the races having been created in separate lands. Most in Darwin's circle however were advocates of unilinear evolution as developed from Lord Kames view of man as beginning as a hunter-gatherer, the state in which European explorers had found among the world's most "primitive" peoples.

Interestingly, many of the earlier evolutionists had not been particularly racist. They saw progress more in universal terms with Christianity, exceptional individuals, and other social factors contributing to man's social development and as have more recent Darwinists such as the late Stephen J. Gould. This included Darwin himself prior to his deconversion. Writing about his experience on the voyage of the Beagle, Darwin praised the Tahitians, whose good behavior Darwin attributed to the work of Christian missionaries. Darwin might have become a theistic evolution had he not lost his Deistic faith due to the problem of evil. Not surprisingly, Darwin's deep racism followed his loss of faith.

Darwin's hagiographers have made much of Darwin's opposition to slavery. Indeed, Darwin was much upset with Lyell who became an apologist for slavery after visiting slave plantations in the American South. But the Darwin and Wedgwood (Darwin's wife) families had been abolitionists since 1807 when Parliament abolished slave trading. How much of their opposition was due to their sincere Christianity and how much to their dislike of competition from merchants in America where slave trading continued is unnecessary to estimate because both causes were a factor.

Slavery or not, native peoples remained in the way of progress in the case of colonial plantations and enterprises in places like India, Australia, South Africa, and China if the labor of natives for dangerous mining and harvesting or their market for industrial products along with tobacco and opium could not be exploited. The need to counter the influence of missionaries who defended the rights of natives explains the rise of racism among ethnologists and anthropologists during the era of Neocolonialism.

In contrast to his attitude towards the Tahitians, Darwin loathed the Fuegians who he met as the *HMS Beagle* crossed through the Straits of Magellan at the southern tip of South America. He particularly despised their dirtiness and looks. He used them as an example of races who are likely to be eliminated according to the '*preservation of the favored races in the struggle for life,'* to use the subtitle of Darwin's *Origin of Species.* Darwin introduced racism into evolution by making evolution a competition between species as well as between individuals.

Social Darwinism has been more often blamed on Herbert Spencer who popularized evolution before Darwin. Spencer's theory of evolution however was based on libertarian economics rather than biology. Thus, Spencer favored Lamarck's theory over Darwin's. Spencer however coined the term *struggle for existence* from reading Darwin's dark explanation of the origin of species. This pleased Darwin who incorporated the idea in the subtitle in his revision of the *Origin of Species.* Alfred Wallace (1823-1913), Darwin's co-discover of evolution by natural selection, who had lived among natives of South America and Southeast Asia, insisted that tribal peoples were no less intelligent and no less capable of progress than white Europeans. Wallace's observations angered Darwin who saw this as challenging their shared theory of evolution by natural selection. In *Descent of Man,* Darwin declared the emotions, behavior, and intelligence as biologically determined according to one's race.

Though presented as a theory of Nature, Darwin had developed his theory by observing the breeding of domesticated animals and by reading Thomas Malthus who declared it impossible to solve hunger by raising the wages of laborers who if not allowed to starve would just have more children. Having established his theory of evolution a law of Nature, this newly "discovered" law of Nature could then be used to justify policies accurately described as Social Darwinist. Not surprisingly, America's robber barons such as Andrew Carnegie and John D. Rockefeller endowed universities that pioneered the teaching of Darwin's evolution in America. Exploiting weak competitors had been found to be but a law of Nature.

Social Darwinist policies could also be used against invalids and imbeciles who required social assistance, as being people who should not be allowed children. The dark science of eugenics was an invention of Darwin's cousin Francis Galton and some of Darwin's own sons. Though ignored in popular accounts of the Scopes Trial, William Jennings Bryan opposed the textbook used by Scopes because it taught Social Darwinism. While serving as Woodrow Wilson's Secretary of State, Bryan had learned from British and French generals the impact of Darwin's theory of evolution on this bloody conflict. Bryan also understood the influence of Darwin's theory in America. On November 18, 1915, Bryan's opponent Clarence Darrow wrote a letter to the *Washington Post* saying 'Chloroform unfit children. Show them the same mercy as shown beasts that are no longer fit to live.' The American eugenics program was admired by Adolph Hitler as the best in the world before the one he established to rid Germany of hundreds of thousands of physically and mentally defective Germans. Apart from the Bible, humanitarian values resort to their horrific pre-Christian state.[25]

Dating the Ice Age and the time since it ended

In current geological terms, the Ice Age is the Pleistocene era, determined to have ended at the beginning of the Holocene, the latter being the most recent geological period lasting up to the present. Together, these two eras form the Quaternary Period, which in geological terms is essentially the same as the era of man.

Prestwich estimated the era of the drift deposits, the same as formerly believed to have been deposited by the Flood, as occurring over about a 20,000 year period. But following his usual way of expanding geological eras, Lyell guessed what was coming to be seen as an Ice Age as being about 220,000 years. Due to not recognizing the diluvium as the product of the Flood, different layers of loess and trees that glacier scientists found in the American and European diluvium deposits seemed to suggest several Ice Ages. But big questions remained. What caused the Ice Age? How long did each of the Ice Ages last? And when did the last Ice Age end?

---

[25] c.f. Tom Holland, *Dominion: How the Christian Revolution Remade the World* (2019, Basic Books). Though an atheist himself, Holland documents how Christianity has changed the entire world to recognize humanitarian values.

James Croll of Scotland and later Milankovitch of Serbia suggested the cause of the Ice Ages as being the axial procession of the earth's tilt. This tilt is responsible for the earth's summers and winters. But 80,000-year precessions which Croll calculated didn't match the geological evidence of the various Ice Ages, while Milankovitch's complex calculations are still debated. Though it doesn't impact the arguments being made here except to point out the infinite plasticity of scientific dating, the era of the Ice Ages was gradually extended, first to 650,000 years by Penck and Brukner in 1909, and more recently to about 2.4 million years. The layers of drift that led to Penck and Brukner postulating multiple Ice Ages are easily explained as shifting directions of erosion from a single worldwide Flood.

The easiest and most reliable of the geological eras to establish should be the most recent, the time since the end of the last Ice Age. Following his father-in-law Horner, Lyell had already calculated the error of civilized man, which would be based on the beginnings of Egyptian civilization and which Lyell had finally concluded could have been no later than about 12,000 years ago. This mythically derived date will have profound consequences for dating prehistoric chronology from the beginning of the Neolithic period to the first historical civilizations in the Near East and Europe.

Sir Arthur Evans used the same rate of deposition to date the first Neolithic settlement in Greece as being between 12,000 and 10,000 BC. But why would the rate of the buildup of silt covering archaeological sites in Greece be the same as those covering sites in the Nile delta? In 1912, Baron de Geer used varves from the meltwaters of glaciers in Sweden to find the end of the Ice Age there at 10,000 years. As also the case of Horner and Lyell's calculations, the numerous highly uncertain assumptions involved in these different measurements don't seem to matter. Understandably, none of them cared to report the numerous measurements that didn't suit them until they were able to adjust their calculations to find some that did. That may explain why, after seventy years of scientific dating, the era of the Holocene still dates to about 12,000 years as Herodotus calculated from Egyptian myth and which Lyell published in his influential 1863 book, *The Geological Evidence of the Antiquity of Man.*

The development of Paleolithic and Neolithic prehistory

The archaeological understanding of 'man in the drift,' soon to be understood as 'the cavemen of the Ice Age,' was put forth by Darwin's neighbor and friend Sir John Lubbock (1834-1913). According to Trigger, Lubbock's *Prehistoric Times as Illustrated by Ancient Remains, and the Manners and Customs of Modern Savages* was almost certainly the most influential work dealing with archaeology published in the nineteenth century.[26] Lubbock focused most of his work on the low state of living peoples. He enjoyed documenting the nastiness of savages, but Lubbock vehemently denied that degeneration of culture had ever taken place. He projected into prehistory how he viewed living "savages," including native Australians who Lubbock and his colleagues like Darwin's bulldog Thomas Huxley believed weren't fully human.

---

[26] Bruce G Trigger, *A History of Archaeological Thought* (1989, Cambridge University Press), p.114

Lubbock determined that those who crafted the unpolished flint weapons found in the drift along the riverbanks had to be the same as those cavemen who were also found under drift deposits. But instead of dating by supposed geological strata, Lubbock shifted to dating man according to a chronology based on an imagined progressive development of artifacts. He believed those found in the drift to be an earlier and less advanced people than those in the mounds who crafted similar rough-hewn tools but who also had a few polished tools. Using this criterion, he distinguished the *Paleolithic* (Old Stone Age) cave men from the *Neolithic* (New Stone Age) men in the mounds. Thus, the Paleolithic era became the same era as the Pleistocene, the Ice Age.

Thinking in Social Darwinist terms, European archaeologists would reckon that their white Europeans ancestors who originated in the Caucasus had killed and displaced the cavemen who occupied Europe during the Ice Age. Lubbock failed to notice that the drift also covered the "Neolithic" mounds, megaliths, and kitchen middens, that he so extensively discussed in *Prehistoric Times.* These mounds were created by the same great Flood as created the drift deposits.

The impossibly long Paleolithic

Viewing these drift deposits as being the product of a long Ice Age instead of a being the deposits of a great Flood lasting about one year meant that the humans using only un-polished tools lived with a primitive but unchanging technology for hundreds of thousands of years. That is an impossible accomplishment for tiny bands of hunter-gatherers who must be constantly on the move as local sources of food are soon depleted. Unlike animals, clothing for warmth and protection doesn't grow on human bodies and in contrast to animals, human babies are slow to mature thus a liability during their early years. How did these tiny bands fail to get wiped out by famine, freezing, flood, fire, drought, pestilence, disease, or else killed either by fierce animals or competitive bands seeking their provisions? They could not have been very bright not to have learned in all that time to domesticate the sheep and goats they were supposed to have followed or plant some of the grains they were gathering. This is all the more astonishing when we consider that paleolithic man was highly skilled in the far more difficult tasks of making fires, clothes, leather sandals, blade-and-burin tools, traveling across the entire planet, and learning to adapt to the most difficult environments as supposed to have existed during the Ice Age. The truth is these bands were simply living on the outskirts of the highly developed civilizations of the pre-Flood world.

Should these Paleolithic bands have lived alone in the earth over such a long period, they would have regularly returned to the most fertile camps, those near springs and rivers. Regardless their primitive technology and the small numbers who could dwell at single sites, after hundreds of thousands of years, the refuse from their meals, shelters, campfires, and industries would have accumulated to a vastly greater depth than the great shell middens that Lubbock assigned to Neolithic man. However, none of the Paleolithic sites show signs of such extensive occupancy.

Edouard Lartet (1801-71), one of Lyell's disciples, learned to date these Paleolithic findings according to what he supposed as the era of flourishing of various now extinct animals whose remains were also discovered in caves throughout Europe. (1) The earliest men found in the drift belonged to the Age of the Cave Bear. (2) Then came the Age of the Mammoth and Wooly Rhinoceros. (3) Next came the Age of the Reindeer. (4) Last, came the Age of the Auroch (wild cattle) and Bison. An even earlier Hippopotamus Age was soon added. Following the methodology of his mentor Lyell, Lartet pulled eras like rabbits from his zoological hat.

Though supposed arranged according to geological periods, Lartet's periods were based on what he saw a progressive development of artifacts. Gabriel de Mortillet (1821-98) continued Lartet's work, though replacing the names of Lartet's zoological eras with the now familiar archaeological ones. The Age of Hippos became the Chellean. The Cave Bear and Mammoth Age became the Mousterian. The Reindeer Age became the Magdalenian. Due to his belief in unilinear cultural evolution, Mortillet searched for signs of whether what appeared as ordinary rocks (eoliths) might instead indicate mankind's first use of tools, which he would then assign to the Lower Paleolithic, even if found lying on the surface of the earth.

Regardless the "Lower," "Middle," and "Upper" classification of the Paleolithic, which reflected the earlier diluvium understanding, strata was hardly ever used for dating the Paleolithic. There is no valid reason for supposing that all these eras were other than concurrent. But the larger skulls of the mega-fauna apes and gibbons gave plenty of candidates for *homo erectus* of a supposed “Lower” Paleolithic. The cavemen were of course found in caves, some using them as temporary hunting shelters (thus modern humans) and the more impoverished, having developed rickets due to these caves being their permanent homes (thus, Neanderthals).

Very unexpectedly, in 1880, a little girl found highly skilled drawings of mammoth and other extinct animals on the walls of caves in southern France occupied by Paleolithic men. Such skillful and beautiful art on the part of primitive cavemen didn't suit the dogma of unilinear evolution. Thus, these drawings weren't recognized as authentic until the end of the nineteenth century when Mortillet at last agreed to assign them to the most recent era of the cave men. This wasn't however the only discovery of the late nineteenth century that was challenging the dogma of unilinear evolution.

Chapter 6

UNDERSTANDING PREHISTORY THROUGH DIFFUSION AND MIGRATIONS

If the Industrial Revolution together with belief in progress inspired the theory of evolution, it was the well-known facts of European and ancient history that informed the beginnings of scientific archaeology as pioneered in the Scandinavian countries. The actual history of man knows nothing of unilinear evolution. Rather, social, political, economic, and technological progress is specific in its origins and spreads to other lands through migrations, trade, missions, and invasions. The latter can also bring civilizations to an end as resulted from the barbarian invasions of Rome and from European invasions of native civilizations in Mexico and Peru.

Material degeneration necessarily accompanies long distance migrations as migrating peoples get isolated from market economies. A shift to a less refined, more rustic style of living happened to the first European settlers of America. Spiritual degeneration may also occur if those who embark on long and dangerous journeys are seeking to gain wealth by capturing slaves or acquiring land by eliminating native peoples. Those who venture to new land often include criminals seeking to escape punishment or else were sent to the new lands as punishment. But a few were law-abiding citizens willing to endure great hardships to enjoy religious and other kinds of freedom while aiming to live peaceably with those who had migrated to the great undeveloped wildernesses long before them.

A similar material degradation happened immediately following the Flood. Due to the great loss of population and consequently specialization in the practical arts, those surviving the Flood faced a more rustic if also more peaceful style of living than existed in the materially refined Early Bronze Age. The art of the fast wheel, the making of painted and stone pottery, the construction of great stone buildings with precision, and likely many other arts were temporarily or permanently lost, though not important ones like writing and spoked-wheel wagons.

The need to maintain unity and order that are essential for survival over long and difficult journeys explain the hierarchical societies of many native peoples where obedience was enforced by torture. The practice of cannibalism as was so often found in the New World is likely the result of the long crossing of the Pacific Ocean. Pre-Christian Europe wasn't much different. Torture and slavery were common practices in the classic civilizations of Greece and Rome as in almost every pre-Christian nation, including some like the Phoenicians and Celtics who also practiced human sacrifice.

Various kinds of degeneration explain the spread of man as seen in the light of the Bible, but a view held only by a few anthropologists such as the Archbishop of Dublin Richard Whatley (1787-1863) and Lyell's Canadian friend John William Dawson. Whatley argued that there is no evidence that savages unaided had ever developed a less barbarous way of life. Thus, it followed that 'humanity must have originally existed in a state far superior to that of modern savages.'

Light spreading from the Orient

Following the path of his Scandinavian predecessors, Oscar Montelius (1843-1921) developed the first relative chronology for European prehistory by seriating the artifacts he found in the museums of European nations. His seriation showed that the peoples of northern Europe had spread from the Orient, as the ancient lands of the Bible were then called. V Gordon Childe (1892-1957), who would further develop Montelius's prehistoric European chronology, would summarize his predecessor's reasoning in 5 propositions:

1. Civilization in the Orient is extremely ancient.
2. Civilization can be diffused.
3. Elements of civilization were in fact diffused from the Orient to Europe.
4. The diffusion of historically dated Oriental types provides a basis for bringing prehistoric Europe within the framework of historical chronology.
5. Prehistoric European cultures are poorer than contemporary European cultures means that civilization is later in Europe than in the East.

Montelius was the first to subdivide the Stone Age, Bronze Age, and Iron Age periods into epochs of Early, Middle, and Late using seriation of styles. He further subdivided these periods by a series of Roman numerals. William F. Albright would later extend these classifications into the Levant of the Ancient Near East.

Montelius's understanding of the light of civilization spreading from the Ancient Near East reflected the work of anthropologists of his day who saw civilization spreading by diffusion as might happen through contacts such as missionaries or trade. Following those who believed that civilization began in Egypt, Grafton Elliot Smith and W.J. Perry taught that the arts of embalming and pyramid building spread even to the Americas, though being increasingly of an inferior, degenerate style. The Americas however have far more similarities with the nations of Far East Asia.[27]

Believing that certain inventions such as the bow and arrow are too complex to have been invented more than once, the ethnologist Francis Ratzel (1844-1901) insisted that progressive development depended on the spread of inventions from superior nations like those of Germany. He aimed to find more living space for the German people, thereby influencing the rise of German imperialism. Ratzel visited places where German immigrants had spread their superior German ways to the United States.

Archaeological *cultures*

As early as 1851, Daniel Wilson, who was the first to use the word *prehistory,* had noted stylistic differences between Iron Age sites in Scotland and those in Scandinavia. Different styles may reflect different peoples rather than stages of development. The term that came to be used for a distinct assemblage of material remains believed to be associated with distinct people groups was archaeological *culture,* as the case of the Khirbet Kerak culture that I mentioned above which spreads from the area of this mountain.

---

[27] Michael D. Coe, *The Maya* (1993, Thames and Hudson), Chapter 2

The idea of diffusion didn't necessarily challenge unilinear evolution, but Ratzel identified the spreading of inventions and progress as being due to a specific and superior culture. His disciple Franz Boas (1858-1942) used Ratzel's concept of culture to develop the first prehistoric chronology for pre-Columbian America. Ratzel and Boas sought to understand archaeology by studying the migration of cultures.

An archaeological culture is the extension to the material remains of ancient people of a term formerly applied to the shared values, manners, religion, government, and laws of a particular class or entire nation. J. G. Herder (1754-1803), who along with Immanuel Kant and J.G. Blumenbach (1752-1840) founded competitive schools of scientific anthropology in Germany, coined the word *culture*, formerly used in agriculture, and applied it to the distinctive literature, religion, and traditions of those sharing a common language. While the racism of Kant and Blumenbach was based on geography, Herder introduced a potentially new source of racism, one based on language. Herder's theory gave rise to the nationalism of the Romantic era that challenged the internationalism of the Enlightenment which Frederick the Great had promoted in Prussia.

Herder had particularly in mind his German language then spoken in numerous independent principalities across Central Europe. Following Napoleon's invasion of Prussia and a famous 1808 speech by J. G. Fichte, German nationalists sought to establish a more prosperous and powerful German nation by uniting the German speaking states. Because of their advocacy, Germany would become the first geopolitical nation created on the basis of a common language since ancient Babel, though it would inspire many other groups to do the same either through peace treaties, terrorism, or war.

Inspiring the creation of a nation based on language requires those speaking such a language to have shared a glorious history. Unfortunately, the written German language was largely the creation of Luther's translation of the Bible. Hence, the German speaking peoples had no written annals by which to teach the members of the new nation their shared and glorious identity. German nationalists however hoped to find the origin and ancient history of those speaking the German language through comparative philology and through the study of German culture including its pagan religion and orally transmitted folk tales. The latter inspired the linguists known as the Brothers Grimm to collect and publish German folk tales.

The idea that history can be reliably preserved through oral means went back to Isaac Newton's teaching about Egypt and Israel's preservation of history prior to the time of David and Solomon when he believed that writing was invented. Newton was influenced by the rabbis who claimed that the teachings of the Talmud had been passed down orally from Moses as the so-called oral Torah. These German nationalists claimed that the same was true of the Zend Avesta, the religious writings of the ancient Iranians, as well as the Vedas of India. But there is no evidence that any of these Eastern religious texts were written down earlier than the Christian centuries. Not surprisingly, the German higher critics supposed the same processes in the composition of the Bible.

Poems and songs are useful for teaching and preserving sagas and traditions in pre-literate societies. But maintaining schools devoted to teaching poems and songs by oral transmission requires the stability of a priestly class which, owing to wars, changes in governments, social and economic positions, can hardly be maintained over such long periods. The reliability of oral transmission can be judged by the wide variations in the accounts of the Flood preserved by the various nations. The oral stories behind the Vedas probably date from Cyrus's conquest of India, while the Hindu religion reflects the class society of the ancient Greeks which dates to the conquest of India by Alexander the Great. My scholarly Indian friends tell me that the oldest temples in India are not those of the Hindus, but those of the Buddhists!

Searching for the origins of nations through language requires the disciplines of philology and comparative linguistics. British philologist William Jones (1746-1794) helped launch such studies by noting the Indian Sanskrit language's close kinship with European languages including Greek, Latin, German, and Celtic. Based on the table of nations in Genesis 10, Jones referred to these as *Japhetic* languages. Not wishing to see the German nation as coming from the obscure Gomer, son of Japheth, son of Noah, and from a religion and literature of Semitic peoples, German nationalists preferred to find the original homeland of those who they called Indo-Germans either in Iran, India, or in one of the classical civilizations such as Greece.[28]

One ancient culture speaking an Indo-German language that German nationalists believed reflected the martial culture of which

---

[28] Tuska Benes, *In Babel's Shadow: Language, Philology, and the Nation in Nineteenth-Century Germany* (2008, Wayne State University Press)

Prussian Germans were very proud was recognized from words pertaining to horsemanship used by the ancient Aryans [Iranians] of Persia who they believed *to* be the same as those who invaded the dark-skinned peoples of India as mentioned in the Vedas. The Sanskrit language in which these Vedas were written is essentially Old Persian which was carried into India by the invasion of King Cyrus who is mentioned in the Bible. But German nationalists aimed to avoid the "Semitic" Bible and discover this history entirely from pagan sources, real or imagined. Thus, Aryan replaced Indo-German. Some German nationalists even hoped to find the roots of German Christianity in the Caucasus, Assyria, or some other "non-Semitic" land!

Cultures such as those speaking the German language could perhaps be associated with distinctive archaeological spreads. Identifying the spread of archaeological cultures with those speaking the German language was an idea based on the German tribes who invaded Western Europe following the fall of the Roman Empire. Using the swastika found on ceramics in the spread of the Aryan race, Gustaf Kossinna (1858-1931) sought the homeland of the Aryan races in the northern part of Germany. He attacked colleagues who sought the roots of German civilization in the Orient or on one of classical nations of southern Europe. V Gordon Childe however was sympathetic to the Marxists of Russia who preferred to see civilization radiating from the Slavic lands such as those nearer the Caucasus.

The Caucasian origins of white people as promoted in Europe and America by Ice Age pioneer Louis Agassiz was popularized in the nineteenth century by the scientifically acclaimed *American school of ethnology.* The American school was based on a wide collection of skulls by the physician Samuel George Morton (1799-1851) but understood according to the teachings of his followers Josiah C Nott and George Gliddon. The American school claimed there existed measurable differences in the skulls of races which indicated differences in intelligence. Arising prior to the teaching of evolution, the racism of the American school was based on the teachings of Isaac LaPeyerre (see pages 40-41), except that it was white people who spread from Noah's Ark whereas the other races were from men before Adam. These ethnologists lamented how a literal reading of the Bible, where all races are seen as having come from Noah's family, was holding back the progress of science. LaPeyerre was compared to Galileo as a persecuted forerunner of science. Though the teachings of LaPeyerre were soon to be replaced by the racist teachings of Darwin, this school's view of the importance of the size and shape of skulls continues to inform modern anthropology's view of the antiquity and evolution of man. Even more did it suit the new Aryan view of the origins of white people.

Leaders of the American school of ethnology translated the writings of the French aristocrat Arthur de Gobineau (1816-1882) whose racist teachings were aimed at supposed differences in Europeans as could be seen in the class system of Europe's *ancien regime*. The German born philologist Max Mueller (1823-1900) taught the Aryan theory in England even if he regretted some of the racist implications. These teachings were much responsible for the invasions, mistreatment, and mass executions of civilians during both World Wars. Both diffusion and racist inspired migrations dominated late nineteenth and early twentieth century anthropology and prehistoric archaeology. But whether racist, nationalistic, or evolutionist, archaeological prehistory was written to accord with the values and views prevalent in the Neo-Colonial powers of the late nineteenth and early twentieth centuries.

After the fall of Hitler, the Aryan languages were renamed Indo-European, as originally suggested by Thomas Young. But philologists and archaeologists continue to search for the original homeland of those speaking the Proto-Indo-European language (PIE), most often finding it either in ancient Anatolia or the Caucasus. These aren't so distant from this mountain where exists the ancient homeland of those speaking all languages.

The post-colonial embrace of neo-evolution

The archaeological work upon which I traced the worldwide migrations back to this mountain depend on the work of archaeologists such as Montelius and V Gordon Childe in Europe and Franz Boas in the Americas who worked from the late nineteenth to the middle of the twentieth century. So do the relative chronologies still being used everywhere today for understanding the prehistory of various countries. Unfortunately, studies of worldwide migrations along with integrated historical understandings of mankind's prehistory have mostly disappeared, replaced by the processual archaeologies based on neo-evolution which presume civilization in most parts of the world developed independently of the Ancient Near East where the Bible reveals as being the birthplace of all mankind.

There were however reasons for this shift. One was the cultural imperialism and racism as I have just discussed that so informed the archaeology of this period, the most dramatic being the archaeology of Aryan invasions based on German nationalism which prepared the way for Hitler. But there were other abuses including bad archaeology in such places as America, Africa, and Australia inspired by racism. Montelius had avoided invasions while seeing the diffusion of culture into Europe radiating from the Orient. Though combining Kossina's concept of culture with Montelius's relative chronology to further develop European prehistory, as a Marxist, V Gordon Childe didn't champion one race above another. But he began his work in archaeology by trying to understand the origins of the Aryan peoples. Thus, archaeologists not wanting to see Europe depending on light from the Orient used him as their chief example to attack the study of archaeology in the light of diffusion or migration.

Another reason was due to not recognizing that most prehistoric peoples lived before the Flood. Archaeologists must work with whatever material remains survived from these ancient civilizations. Thus, there was a turn to neo-evolution which differed from the older evolution where civilization might have been advanced by special gifted individuals or spiritual sources as in the case of theistic evolution. According to neo-evolution mankind's future is fully determined by his physical environment. This meant completely rejecting any historical sources for the rise of civilizations as the case of light coming from the lands of the Bible.

The strongest case for agriculture and civilization having developed independently of the Ancient Near East was in America where the domesticated plants were almost entirely different from those in the Old World and where the origin of man in America was believed to have dated from the Paleolithic cultures that crossed from the Old World to America by way of a temporary land bridge arising from falling oceans during "the last glacial maximum." The idea that native Americans had crossed from Asia via a land bridge however is far older than the Ice Age theory. It dates to the speculations of the earliest European explorers of North America.

Europeans didn't know that Far East Asians had long been superior mariners. Wooden boats do not long survive, but archaeologists should have learned from anthropologists that the most "primitive" peoples knew how to fish and build boats from the time that Europeans discovered them. Should we not expect children of Noah to have retained maritime skills? The earliest settlements being along the Pacific coast makes it clear that that America was first settled by fishermen. Before and after the Flood, the first American settlements were in Ecuador. [29] Why Ecuador? Experienced sailors know the danger of lost ships sailing in circles. Thus, the safest way for lost mariners to avoid sailing in circles is to follow the sun. No matter where in the Pacific one begins his voyage, by following the sun one will eventually arrive in Ecuador. The Formative Culture for native American civilization begins in Ecuador.[30]

Mistaken belief in the native American civilizations' supposed independence of Old World Archaeology was not the only reason that Americans took the lead in what by the sixties was being called the *New Archaeology.* This trend was led by archaeologists inspired by the invention of radiocarbon dating, by statistical analysis of artifacts, by the paleobotanists' study of ancient diets recovered from archaeological sites, and by using computers in the service of archaeology! For some, archaeology seemed on the verge of becoming a science instead of one of the humanities. Science of course is very useful for archaeology. No archaeological project

---

[29] Betty J. Meggers, Clifford Evans, and Emilio Estrado, *Early Formative Period in America: the Valdiva and Machalio Phases* (1965)

[30] James A. Ford, *A Comparison of the Formative Cultures of America* (1969)

needs these and many new technologies more than the one our scientists hope to use for the study of this mountain and region.

Advocates of what was called the *New Archaeology* supposed that archaeological sites and rise of civilizations might be explained according to General System Theory. This was in line with the scientism that the American social sciences had been promoting since the end of the nineteenth century which aimed to explain America's exceptionalism as due to its leadership in science.[31] As a new nation, America lacked the long history and the numerous historical places and markers that so informed Old World archaeology.

As a systems scientist and pioneer of the world's first distributed digital and fiber optic networks, General Systems Theory was also my field. I used a systems understanding to automate the world's most complex infrastructure networks. But I knew that the key to developing innovative solutions was my unique understanding of telecommunications and infrastructure *history.* I had also to know the specific needs of the various rural and urban areas for which I was designing these systems, along with the operations of various kinds of companies and government agencies who served them. To excel in engineering, one must be interested in human beings, the opposite of the stereotype of engineers as technical nerds. Good engineering required more than a knowledge of technology, but also knowledge of the unique histories of the places and unique needs of those needing telecommunications services. I was

---

[31] c.f. Dorothy Ross, *The Origins of American Social Science* (1991, Cambridge University Press)

astounded to learn that archaeologists supposed they could understand humans apart from the light of their specific histories.

Such systems thinking that has been employed to explain the rise and collapse of the great pre-Flood civilizations. Explanations using abstractions such as *complexity* and *collapse* merely restate the problem. Thinking in terms of abstractions masks boring and useless gobbledygook. Archaeology however wasn't the only field traditionally belonging to the humanities that scientism was destroying. *Scientification* based on subjective abstractions so as to follow fashionable trends was undermining higher education in America. Such teachings were often accompanied by the idea of a progressive scientific elite who should run society. Sad to say, as an honor-class engineering student in in the late sixties, these teachings appealed to me.

Ironically, that means that the earlier racist inspired migrations upon which prehistoric chronologies are based must continue to interpret archaeological remains in the processual understanding of archaeological prehistory. But migrations are a fact of history and need not be interpreted in racist ways. Nothing is less racist but instead a solution to racism in recognizing the people of all nations as belonging to a single family who lived less than five thousand years ago and who began their migration to all parts of the earth from this very mountain where we are gathered. Considering that Noah had three daughter-in-law's who descended neither from himself nor his wife, we should not be surprised that the peoples of the world reflect those three genetic divisions while those living closest to this mountain reflect them all.

Chapter 7

ESTABLISHING HISTORIC AND PREHISTORIC CHRONOLOGIES FOR THE ANCIENT NEAR EAST

As Montelius and Childe sought to establish a relative chronology for prehistoric Europe, scholars and archaeologists working in the lands of the Bible had been seeking both historical and prehistorical chronologies for what Childe called the *Most Ancient East*, now referred to as the *Ancient Near East* (ANE). The Ancient Near East is essentially the lands mentioned in the Bible. (The term *Middle East* is the location of a modern military command.) Unlike barbarian Europe or even the classic civilizations of Greece and Rome, literacy had long been present in the lands of the Bible. As Western nations obtained access to lands of the Ancient Near East beginning with Napoleon's invasion of Egypt, archaeologists found monuments, inscriptions, and clay texts with the names of the same Assyrian and Babylonian kings known from the Bible, some even referencing kings of Israel. Explorers found trilingual inscriptions as those on Egypt's Rosetta Stone and the Behistun inscription in nearby Kurdish Iran by which they learned to read long extinct languages including Coptic and hieratic Egyptian, Persian, Elamite, and Akkadian. From the latter they learned to read the Sumerian and Hurrian languages.

Using names and king lists written on monuments and ancient papyrus from Egypt and on clay tablets in Mesopotamia, scholars were able to construct more extensive king lists by which to establish historical chronologies from even more ancient times. Some of these texts mentioned a Flood like the one in the Bible, including kings who lived beyond ordinary lifespans, the earliest of whom lived before the Flood. Working from these, scholars established a historical chronology for the earliest dynasties back to about 3100 BC in Egypt and 2800BC in Mesopotamia. One should however keep in mind that due to gaps in these king lists, the earliest reliable historical chronology dates only from the beginnings of the Neo-Assyrian empires in the late second millennium BC. The much-discussed chronology of Egypt, including the proposed dates for the Exodus, is no exception.

Flinders Petrie (1843-1942), who saw civilization in Egypt as coming from an invasion of Africa by a superior race from the North, was chiefly responsible both for developing an Ancient Near Eastern historical chronology and for connecting Montelius's prehistoric European chronology to the Ancient Near East. Working with Arthur Evans, he connected his historical chronology of Egypt to those of the Aegean, while Mortillet connected the Iron Age Hallstatt and La Tene cultures to the historical Celts who would eventually inhabit Britain. For the prehistoric eras of the Ancient Near East however the archaeologists could seek only relative chronologies. Following Montellius, Petrie sought to accomplish this through a seriation of pottery based on his proposal for a supposed development of pottery styles. But the lack of stratified sites in Egypt made it difficult for Petrie to establish a useful relative chronology for prehistoric Egypt. Thus, he began working in Palestine where prehistoric levels there could be linked back to those in Egypt.

How the civilization deeply buried by the Flood confused ANE chronology

Digging through the ruins of ancient Nineveh and in sites extending into lower Mesopotamia, archaeologists found a series of pottery which they wrongly supposed as having been produced in a succession of prehistoric eras. They named these for various sites where the pottery had first been identified. These ranged from Halaf, Hassuna, and Samarra nearer the top of the Fertile Crescent to distinctive kinds of pottery found at the sites of Ubaid, Uruk, and Jemdet Nasser in the south. These are followed by Early Dynasty remains, the first historical period of the Sumerians which according to the current historical chronology began around 2800BC.

As I mentioned above, aside from sites in the highlands, Early Bronze Age ruins found under tels in the great river valleys were too deeply buried by the Flood for easy excavation. The same is necessarily true of everything supposed earlier. In comparison to the upper layers where were found the historically dated Sumerian and Akkadian remains, these deeper remains are massive. It is because these are the remains of what appears a highly populated, highly developed, and materially refined civilization that existed in Mesopotamia prior to the Flood. Many will be surprised to learn that the remains from the great Empire of Sargon the Great, which is reckoned as being the high point of Ancient Near Eastern civilization, are comparatively meager but near the surface and thus easy for archaeologists to recover.[32] The population of the Ancient Near East had profoundly dwindled due to the great and then recent Flood.

Some have wondered why Noah didn't re-use the wood from the Ark. Dismantling the Ark and carrying these heavy timbers down this high mountain would have been an unnecessary burden because there existed plenty of worked wood from the dwellings of those who perished in the Flood. This wood however soon rotted and burned. Pottery, on the other hand, is more durable. It seems that the series of beautiful pottery including that found at Halaf, Hassunna, and Samarra that were produced during the refined preFlood civilization were mined by the survivors of the Flood as they traveled from the land of Arrata along the Tigris and Euphrates Rivers. This pottery spreads in the same general area as that of Late Uruk which, as I mentioned above, is likely the site of the Tower of Babel. Contrary to the present chronology, Late Uruk is later than Jemdet Nasser.

The post-Flood remains seem to begin in what is now regarded as Early Dynasty IIIB remains. Those beneath the Early Dynasty IIIB remains are for the most part contemporaneous. It now seems possible that none of these ruins including those buried by the Flood date much earlier than about 4000BC.

---

[32] Harriet Crawford, *Sumer and the Sumerians* (1991, Cambridge University Press), p. 16

The same is true in Egypt, where the Pyramids built during the earlier refined but violent civilization were likely for protecting treasures. This explains why the Egyptians had no memory of building the Pyramids which these historical Egyptians remodeled as tombs for Egyptian kings. Due to the Flood, the Archaic Egyptian language for which the hieroglyphic script was invented has never been deciphered, a matter that popular histories of ancient Egypt seem blissfully unaware. Neither are they aware that Egypt's Old Kingdom, Middle Kingdom, and New Kingdoms are but modern inventions imposed on the history of ancient Egypt. All these Egyptian dynasties however date from after the Flood. Evidence of the First Dynasty of Egypt was found at the site of Khirbet Kerak, just as we might expect due to the historic Egyptians having migrated from this mountain.

Neolithic Revolutions occurring at the end of the Ice Age

We have next to consider the absolute archaeological chronology for the beginning of the Neolithic in the Ancient Near East which as in every place else in the world remains anchored to Lyell's magical 12,000 bp 'end of the Ice Age.' In 1914, the Egyptologist James Breasted (1865-1935), who founded the Oriental Institute of Chicago, coined the term *Fertile Crescent* to describe the crescent shaped region running from Israel in the south to Mesopotamia in the East where he judged to be the place where agriculture and civilization began. In 1935, V Gordon Childe put forth the notion that successive agriculture and urban revolutions had occurred some place in the Fertile Crescent immediately following the end of the Ice Age. As a Marxist, Childe thought in terms of *revolutions.* According to Childe, the first revolution was when mankind suddenly changed from EpiPaleolithic hunter gatherers to Neolithic agriculturalists. His second revolution occurred after another few thousand years when Neolithic agriculturalist villagers learned to make pottery, forge metal, and build the world's first urban civilization where they would invent stamped seals and writing chiefly for the purposes of accounting for their stores and commercial transactions.

Why and how did Childe suppose that after hunting and scavenging for hundreds of thousands of years, mankind should suddenly recognize the advantages of cultivating the grasses he was reaping and herding the flocks of wild sheep and goats that he followed? Following an idea put forth by Raphael Pumpelly in 1908, Childe suggested that it was because the Ancient Near East was becoming more arid during the colder Younger Dryas era supposed to have followed the Ice Age. This caused hunter-gatherers to spend more time at oases near the springs of rivers where plants and animals would be concentrated. The oases turned into villages where agriculture would begin. I suppose these oases could be seen as little gardens of Eden. But unlike the sons of Adam, it appears to have taken evolutionary man another 2,000 years to recognize that the wild grains they were gathering were becoming more productive and the advantages that could be gained by sowing some of the wild grains.

Dorothy Garrod (1892-1968) managed to link the Paleolithic sites in Europe with the Neolithic sites in the Ancient Near East by exploring Neanderthal burials found in Gibraltar on the coast of Spain, in cave shelters in the foothills of the Zagros Mountains and on Mount Carmel near the Mediterranean coast of Palestine. As was the case in Europe, she found the Near Eastern Paleolithic sites in rock shelters and caves on the sides of mountains, just the places where hunters might want to stay when away from their homes and villages. She connected these Epipaleolithic burials in the Schuqba Cave on Mount Carmel to nearby Wadi Natuf, where she discovered stone sickles and what has become accepted as the beginning of world agriculture.

The excessive antiquity of prehistoric man in the Ancient Near East due to the magical 12,000 bp 'end of the Ice Age' vastly stretched the antiquity of the ANE cities. Together with the fact that Charles Lyell taught archaeologists to equate depth-of-deposit with time, this greatly impacted the chronology of individual sites. Consider what this means if a 200-year-old city became entirely covered by the deposits of the Flood of a single year, thus creating an Ancient Near Eastern tel. To make sense of these artificially created long-period sites, archaeologists are prone to excessively sub-divide the prehistoric chronological levels of sites that have neither been destroyed nor substantially rebuilt. Such attention to artificial periods averts attention from what was likely to have been a brief but rapid development of these sites and cities prior to the Flood. This is true of prehistoric archaeological sites all over the world but especially in the Ancient Near East as the case of those searching for the world's oldest city.

Discovering 'the world's first and oldest city'

Robert Braidwood (1907-2003) began working with Breasted in the foothills of the Zagros Mountains hoping to discover the world's oldest city. Dame Kathleen Kenyon (1906-1978) then working in Palestine at the site of ancient Jericho had the same aim. Being near the center of the Fertile Crescent and the world's oldest civilization in Mesopotamia, it was widely anticipated that Braidwood's site of Jarmo was the most likely candidate.

Following John Garstang who began the excavations at Jericho, Kenyon reasonably determined that the spring coming from the bedrock that created the oasis was the reason that Jericho became an early settlement. She found at the site evidence of the same kind of agriculture culture as Dorothy Garrod had found at Wadi-Natuf. That put the beginnings of the first temporary village at Jericho to perhaps 10,000 BC. Beginning around 8500BC according to the current chronology and what Kenyon called the Pre-Pottery Neolithic A period (PPNA) were the first houses and a massive stone wall. She reckoned that the great defensive tower built around the spring and the finely plastered houses must have dated from the same time. In the following PPNB period, the occupants built rectilinear houses and learned the art of plastered skulls.

According to Kenyon, everything wonderfully advanced about the city including plastered walls as fine as that in some modern hospitals dated to the Jericho's initial phase and long before(!) Jericho's pottery. But pottery was the one thing that could establish the date of the city relative to other Ancient Near Eastern sites and would have greatly lowered the date of the early stages of Jericho. In fact, there were plentiful amounts of pottery on the floors of these finely built houses, but Kenyon attributed every one of these pots to intrusions from above dug by later occupants. According to Kenyon, the first to use pottery in Jericho must have lived in tents because Kenyon could not find their houses. They left no other trace of their occupancy for thousands of years until the end of the Early Bronze Age when the city was suddenly abandoned.

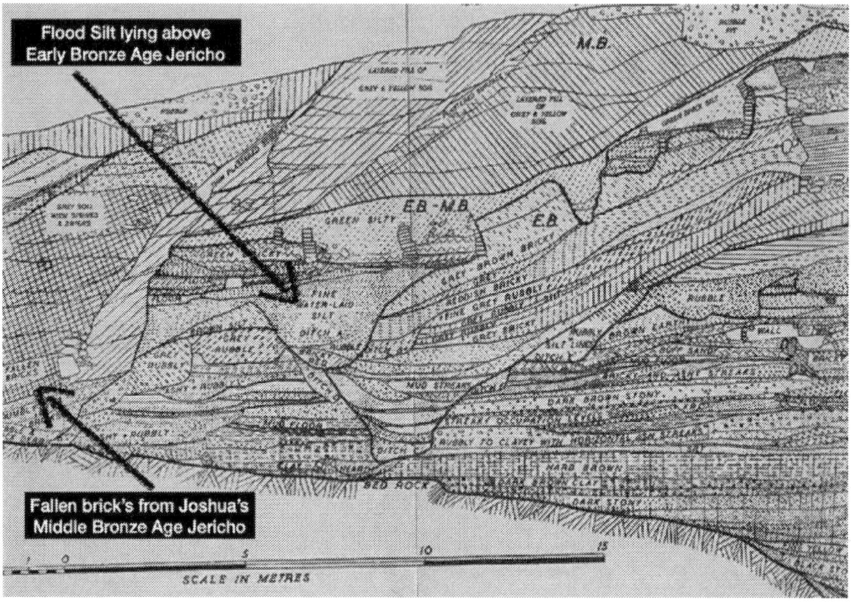

Interestingly, Kenyon's famous stratigraphic drawing (see above) of the long trench which she dug into the Tel es-Sultan mound of ancient Jericho shows 'water-laid silt' lying just above the Early Bronze Age city.[33] To her credit, Kenyon writes a great deal about the great erosion occurring throughout Palestine at just this time. This is something that exists the world over but largely ignored by archaeologists, though not by geologists who discover all across the world at about this time a great number of "local Floods." Among these are the Black Sea Flood, the so-called Missoula floods, the Grand Canyon Flood, the Flood in Tibet and other places in the world supposed due to such things as the breaking of water barriers

---

[33] Kathleen Kenyon, *Digging Up Jericho (1954),* Figure 4

or natural dams, which are but the many local evidences of a single worldwide Flood.

Important to our project, the next occupants of the abandoned Early Bronze Age city brought with them the Khirbet Kerak pottery that streams into southern Palestine from these mountains of Ararat.

William F. Albright, who created the archaeological chronology of the ancient lands of the Bible, as well as his assistant George Wright noted Kenyon's tendency to separate pottery from strata at her other sites such as Tel Megiddo. Kenyon's mentor Sir Mortimer Wheeler knew no place on earth where there existed more sins in the name of archaeology than in Palestine. Wheeler probably had in mind his own student. Archaeologists lamented the fact that Kenyon never published her archaeological excavation of Jericho. But something happened that caused criticism of Kenyon to cease and for archaeologists working in the Levant to use Kenyon's Jericho prehistoric chronology to date their own sites.

Chapter 8

ARCHAEOLOGY'S RADIOCARBON REVOLUTION

Excited about the prospects of finding the world's oldest city, Robert Braidwood formed one of the first multi-disciplinarian scientific consortiums to study the site of Jarmo in the foothills of the Zagros mountains of northern Iraq. This is much as our team of archaeologists and scientists are planning for this region of Mount Ararat. Learning of the recent discovery of radiocarbon dating by his University of Chicago colleague William Libby, Braidwood invited him to use his new technology to date the site. The radiocarbon dates of 4800-4600BC which Libby found in Jarmo were not so different from what Braidwood expected. Libby then decided to date charcoal from the site of Kenyon's Jericho. Charcoal from the city's earliest remains dated from 6250-5850BC. That's when Braidwood ceased getting attention for his efforts to find the world's oldest city. It had already been found.

It seems that no one then thought about the possibility of old carbon in the city that served as the ancient center of the bitumen trade. More significantly, the deep soil covering earliest Jericho which licensed Kenyon to assign long periods for PrePottery Jericho A and B belongs to the same diluvium that defines the entire Ice

Age. It is also the same diluvium that created most of the Ancient Near Eastern tels and which buried the inhabitants of Jericho in the floors of their houses and on the stairs of the defensive tower that also served as Jericho's granary. Due to the presence of these burials, instead of finding "the world's oldest city" Kenyon could have advertised Jericho as "the world's oldest temple."

The world's oldest temple

Kenyon's work at Jericho did however help date the now recognized "world's oldest temple," Gobekli Tepe in nearby southeastern Turkey, a site being excavated by Klaus Schmidt. His method of dating is much like Kenyon's, that is, it is based on what he did *not* find at Gobekli Tepe, which is pottery. If he does find pottery, might it not be judged as intrusive or from a later era as the case of Kenyon's pre-pottery Jericho? Thus, Schmidt dates this site to Kenyon's Pre-Pottery Neolithic Age (PPNA), which is between 9600-8,000 BC.

What seems to have made Gobekli Tepe a better candidate for the world's oldest temple is, on the one hand, the lack of evidence for dwelling activities such as cooking and, on the other hand, the puzzling stone pillars, and animal statuettes, some of which appear to have been in the process of being carved. Some of you gathered here have seen similar strange animal carvings at various places around this mountain which would have better survived burial by topsoil washed down by the Flood. We may now know from whence they come; from sites like this if not from Gobekli Tepe itself. The remote location of Gobekli Tepe, away from sources of water which would be needed for permanent residence, as well as the fine limestone on the hill in which this site sits, points to Gobekli Tepe as being the quarry for these statuettes and limestone pillars. Some are in an incomplete state, as the case of numerous other factories and stores at sites ended by the same Flood that buried this remarkable site.

The only thing that we can be certain that religious interpretations of archaeological sites tell us concerns the minds and thinking of modern anthropologists and how they influenced archaeologists. The sociologist Auguste Comte (1798-1857) and especially the anthropologist Edward Tylor (1823 -1917) imagined an excessively religious prehistoric world. This served in a big way to blind archaeologists to the manifest evidence of the Flood. Emile Durkheim (1858-1917) taught them that religion is something that can be understood scientifically! I can't decide which did more to blind our modern world to the truth concerning origins. Was it their aim to explain everything according to uniformitarian naturalistic science? Or was the religious imaginations of scientists? Absent relevant texts, religious interpretations should be altogether banned from the science of archaeology. Whatever the case, archaeologists will no longer be able to look at prehistoric burials without thinking about the Flood.

How the Flood impacts dates for archaeological prehistory

Klaus Schmidt has confirmed the PPNA dating of Gobekli Tepe with several radiocarbon dates, some from organic material clinging to these pillars, all which date from 9676-6749BC. It would be interesting to learn the radiocarbon dates of soil clinging to the foundation of some of our modern houses. Especially in old houses, many of our yards may have the same kind of topsoil that covered

Gobekli Tepe at the time of the Flood. Across the earth will be found soil loosen by the Flood which contain organic remains such as decaying vegetation and insects, plus charcoal from the burning of ancient trees perhaps tens of thousands of years prior to the Flood. Thus, a great radiocarbon disruption occurred at the time of the Flood. Even if one ignores the Flood, it remains the case that soil with organic remains has been building up for as long as vegetation has been growing and as long as worms, insects, beetles, snails, and other animals have been dying.

Should the bones discovered at a site belong to humans or animals with a diet of ocean fish, the old carbon from these marine diets gives dates far too old. Thus, we should not be surprised that some of the "oldest" sites in Australia, Africa, and America are found along coasts where fish are a substantial portion of diets. Archaeologists know the limits of radiocarbon dating, but it is still generally reckoned as reliable for things as old as 30,000-50,000 years. Perhaps the dating is accurate, but just what is being dated?

Old carbon should not ordinarily be such a problem when directly dating either bones or organic *artifacts*. But other than a few hard bones or antlers which many sites lack, what organic artifacts survive water, decay, insects, and fire for more than 4,000 years? The only artifacts that survive so long have been buried in a desert or inside caves where there is no moisture and where it never rains, or else in oxygen depleted peat bogs or lakes. Other rare opportunities exist as in the case of mammoths buried in the Arctic or Otzi the Iceman found in the Alps. Most of all, the ancient wooden structure reported to be buried under a glacier on the top of this mountain may open a great window to everyday life in the prehistoric world as never before, but which we should now understand as the pre-Flood world.

Another thing that may impact radiocarbon dates older than the time of the Flood concerns the matter of why calibrated radiocarbon dates begin matching uncalibrated dates for years later than about 1500 BC, as if a new normal for radiocarbon in the atmosphere is being reached! All scientific measurements presume uniformitarian conditions, but tree-ring calibration shows this not to be the case for radiocarbon dating. As I will explain in the next chapter, the atmosphere prior to the Flood was drier and colder. How such things impact radiocarbon dating is at least for the present an important but unknown factor.

Such considerations explain why the radiocarbon revolution shocked the world's top archaeologists by so pushing back what were then archaeologically determined dates. After discovering the need for tree ring calibration, the absolute dates of prehistory were just pushed back even further. Due to the need for funding, most of it requiring private sources, archaeologists depend on interest and attention to their work. That comes from excavating "the world's oldest city," "the world's oldest temple," or the world's first this or first that. In *Archaeological Evidence*, I predicted that absolute archaeological dates will be pushed back towards infinity just as we have observed from the beginning of radiocarbon dating.

This remains the case even if many new ways of scientific dating have been developed. The problem with all scientific dating is the absence of a reliable chronometer by which to calibrate these new methods. The problems are not just because of the challenges that dating laboratories face because choosing dates and interpreting sites has been the privilege of the archaeologist licensed to investigate sites, few of whom have expertise in scientific dating sufficient to make good judgements about scientific dating.

The Dead Sea Scrolls remains the outstanding example of the accuracy of radiocarbon dating because we can use paleography to date these scrolls very accurately to calendrical dates. So important are these that Dr. Stephen Pfann (see page 10, above), a paleographer and scientist on the Dead Sea Scrolls team, tests the precision of the world's foremost radiocarbon dating laboratories.

Beyond this, most kinds of scientific dating are calibrated to what we believe is the most accurate chronometer, tree ring calibrated radiocarbon dates. But contrary to popular accounts, tree-rings sequences have gaps or ways of counting rings in which some faith is required, if however not as ridiculous as we have seen in the case of dating the Holocene. To learn the circular reasoning involved in wiggle-path matching of tree-rings, one must learn the scientific argot sufficient to read the calibration issue of the journal *Radiocarbon.* No one understands that scientific dating is as much art as objective science as those most expert in these fields. I look forward to the best of these laboratories using the ancient wooden structure on this mountain to advance the art of scientific dating.

How recognizing the 2300 BC Flood event can clarify archaeological prehistory

This isn't to say that we can't see the Flood in the current radiocarbon dates. The Flood is surely responsible for what archaeologists now call the 4.2 ka cal. B.P. event.[34] As I reported in the *Archaeological Evidence*, I was able to date the Flood to about 2345 BC based on radiocarbon dates from the Jomon sites in Japan, the Harrappa sites in Pakistan, and the Stonehenge structure in England. That's about the same date for the Flood as Archbishop Ussher calculated from the Bible.[35]

Archaeologists are rightly prone to reject dates that make no archaeological sense. But archaeological sense depends on prehistoric relative chronologies. Apart from the problems in absolute dating, not recognizing the impact of the Flood around 2300 BC has created a not insurmountable problem with the relative chronologies of prehistory, a few which I have already mentioned. This is due to the vast number of burial sites created by the Flood and too few occupational sites in archaeological prehistory to which these graves can be assigned. I discussed the stretching of archaeological prehistory from the Middle Paleolithic to the Early Bronze Age. But the burials are so vast that many of the Flood burials are assigned to occupational sites far after the Flood as the case of the Adena and Hopewell burials in Ohio. Other Flood burials have been assigned to their own archaeological periods so as to account for a stretched out occupancy as Kathlyn Kenyon did with Flood burials in the vicinity and hills around Jericho.

---

[34] Bernhard Weninger, 'A Gap in the Early Bronze Age Pottery Sequence at Troy Dating to the Time of the 4.2 ka cal. B.P. event,' *The Late Third Millennium in the Ancient Near East Chronology, C14, and Climate Change*, edited by Felix Hoflmayer (2017), the University of Chicago Oriental Institute Seminars, Number 11, p. 429

[35] c.f. *Archaeological Evidence,* pp 173-175

The good news is that all this can be resolved simply by using the new paradigm of *Scripture Science* as I have been doing here, clarifying the origins of man more than the new James Webb telescope can clarify the origin of our universe. Flood graves all over the world serve as an anchor for both relative and absolute dating of what has formerly been understood as prehistoric archaeology. We may be able to eliminate prehistoric archaeology, replacing it with an historical understanding from the first appearance of man which may have been no more than about 6,000 years ago!

Chapter 9

SHIFTING TO THE NEW PARADIGM

However unknown the source documents that may have been used for its composition, what can only have been a divinely inspired historically accurate narrative of the beginning of man is a powerful light for transforming the historical sciences' currently fragmented, complicated, and muddled account of man's past into an understanding that is clear, integrated, and simple. This ancient light can predict a host of phenomena far more precisely than can the current historical sciences, which is just what is expected of the best science. This light must however be consistently and holistically employed in all the interdependent sciences as in the way I have explained. This radical transformation of the historical sciences becomes a new school of origins which this author refers to as *Scripture Science* to distinguish it from what is known as *Creation Science,* whether the latter be of the young or old earth varieties.

To review, mainstream science's fragmented, complicated, and muddled understanding of mankind's origins stems from its flawed foundation having been built on three great blunders chiefly engineered by Charles Lyell:

1. Re-envisioning what early nineteenth century geologists had correctly understood as the rapid deposits of a worldwide Flood as being instead the gradual deposits of what would come to be reckoned as a series of Ice Ages lasting for hundreds of thousands, eventually to be extended to several million years.
2. Pushing the beginnings of civilization in Egypt, and thus the Neolithic period, from less than 6,000 back to 12,000 years.
3. Presuming that humans found living with animals now extinct in Europe predated Europe's first farms, villages, and forts, remains now classified as either Neolithic or Early Bronze Age.

This resulted in pushing the date of the remains of men found in the diluvium deposits, now supposed as from the Ice Age, from about 4300-6000 years to hundreds of thousands of years and viewing some of them as subhuman cavemen rather than impoverished cave dwellers and the temporary camps of hunters living in the Early Bronze Age. This artificial stretching of mankind's most ancient history resulted in an impossibly static era of mankind's history for hundreds of thousands of years followed by sudden revolutions at the "end of the Ice Age." Contrary to what is taught in both mainstream and Creation Science, there was never an Ice Age. What is now understood as deposits from a supposed Ice Age were from the same Flood that covered prehistoric burials in low-lying lands all across the earth.

As noted in the Introduction (page 5), most of us are not conscious of the foundation principles of the paradigm by which we have been taught to understand evidence. The same is true of those who have been taught to read the Bible according to Creation Science. Despite widespread belief of it being based on a literal reading of the

Scriptures, today's young earth Creation Science has more in common with the instantaneous creation of a fully functioning world with the appearance of age as taught by the Deists who immediately preceded the young earth Creationists. This essentially removes from the Creation account the historical understanding that is unique to the Bible and essential for the creation of today's historical sciences. Note however that the mostly fish-killing Flood of the Creationists is hardly the one described in the Bible! Neither can these Creationists explain why rivers (Tigris and Euphrates) and lands (Cush and Havilah) of the antediluvian world survived their earth-destroying young earth Flood. Thus, Creationists can neither find the vast worldwide burials and civilizations of those who drowned in the Flood nor trace the migration of the earth's present inhabitants from these mountains of Ararat.

Traditional readings however appear to be literal readings if generations have been taught to read and either defend or attack the Bible in such a way. They will not be seen for what they are, *scientific* interpretations read onto the Scriptures *(eisegesis* rather than *exegesis).* The Bible does not literally tell us that the days of Genesis were of 24-hour duration. The young earth view assumes the non-transcendent understanding of Absolute Time that Isaac Newton employed for his theories of classical physics. Newton called his Absolute Space and Absolute Time the 'sensorium of God,' as if God were a physical rather than spiritual being. This was a scientific view imposed on the text of the Bible. Not surprisingly, discussions concerning the duration of the days of Genesis began immediately following the publication of Newton's *Mathematical Principles of Natural Philosophy.* But as Exodus 20 points out, the 6 days of Creation were God's days and not man's days. The Scriptures do literally tell us that the duration of God's days is not the same as that of man's days.[36] Though having beginning and end, a 'before and after,' God's days are not determined by an external marker of time as is the case of man's days. Not until the fourth day of Creation did God created the sun to mark the duration of man's days, which the scriptures state as serving that purpose.[37] This defense of a young earth supposed entirely based on the light of the Bible is well compared to when the Bible appeared to teach an earth-centered universe due to the scriptures having long being interpreted according to Aristotelian Science. Especially if we are unaware of scientific beliefs that we unconsciously hold, they will impact how we read the Bible. As the name suggests, *Scripture Science* employs a more literal view of the Bible than does young earth *Creation Science.*

Young earth Creationists however have honestly and rightly contended against most Old Earth Creationists that the Bible clearly teaches a worldwide Flood. I am indebted to the young earth Creationists for making this clear to me. Until now, the only school of origins teaching a worldwide Flood was that of the young earth Creationists. This probably explains why most young earth Creationists have a higher view of the Scriptures and a more skeptical view of science than do most Old Earth Creationists. They need however to be more skeptical of Creation Science.

We often hear, 'The Bible isn't a science text.' This is true, though not in the way those who say this generally mean. However

---

[36] Psalm 90:4, 2 Peter 3:8

[37] Genesis 1:14

inconsistently, these same scholars also tell us that the Bible teaches or presumes an ancient science. The Scriptures do claim to be an historical text giving us the truth about the past which is vital to a good scientific understanding. But while truth does not change, scientific understandings do.

Note that the Scriptures address God's Creation in a concrete and historical way, referencing the specific earth, waters, sun, moon, and sky that the people of every nation on earth are familiar, visible things which we refer to using *proper nouns.* Science, on the other hand, concerns the description and functioning of ordinary things and abstract concepts of our own invention which we describe in our language using *common nouns.* The Bible recognizes the value of science, usually translated as *knowledge* or *wisdom.* New theories appear to address the little that science can physically observe and partially understand as knowledge of the Creation increases.

Instead of teaching from the light of the ever changing sciences, which are indeed culturally dependent, the wisdom taught in the Bible concerns the eternal truths about the Creation, about us, and about our Creator. Thus, while science is culturally dependent, the Scriptures aren't. From the beginning of Christianity, the same biblical truths have been taught, understood, and applied by people of every nation, language, and culture.

A few matters relating to the current and new paradigms need addressing.

Concerning the other historical sciences

In this monograph, I have focused on the archaeological and geological sciences. Other historical sciences impacted by the third millennium BC Flood are beyond the scope of this summary of archaeological prehistory and recent earth geology. Some questions such as the change of fauna on every continent due to this event, the likely migration routes of animals after the Flood, the capacity of the Ark for all the kinds of animals, the current species that derive from these animals, and the sources and destination of the Flood waters, I briefly covered in the *Archaeological Evidence.*

Some have noted the confirmation of the claim in my 2011 book that more water is to be found under the continents than in all the earth's oceans. In 2014, we learned there exists more water under North America alone than in all the oceans. Most of this water exists in a crystalline state. But water is lighter than molten rock. So how did this water get so deep inside the earth? Following the narrative of the sacred text,[38] my 2011 book explained how water in a plasma state as must exist even deeper inside the earth is heavier than the molten and solid rocks that cap our planet.

There is also the matter of the ice sheets in Greenland and Antarctica which this wholesale rejection of the Ice Age suggests as being of recent origin. Based on WWII aircraft found a few decades ago under a deep layer of ice near the coast of Greenland, I suspect that ice accumulates far more rapidly than currently supposed even

[38] Genesis 1: 1-2, 6-7; 2 Peter 3:5

in the driest parts of these polar land masses. Contrary to claims, annual layers cannot be counted in the deeper layers where the compacted snow has fused into ice. But the remarkably even growth rings reported to exist in the ancient wooden structure on this mountain suggests that the axial processions of the earth's tilt that are responsible for our summers and winters, the variations of which many have seen as being the cause of the "Ice Age," may have begun as recently as about 2300BC.

This suggests a different method of watering the earth prior to this event which may explain why what are currently known as large tropical animals were then living in the arctic regions and why great civilizations were then living in the desert belts that are created by the earth's current hydrological cycle. Contrary to the teachings of young earth Creationists, the world before it ever rained(!) was cooler and drier. We know this by the consistently larger size of the fauna, helping these animals keep warm and hold their moisture. But the temperature variations between summer and winter before the Flood were not so extreme as after the Flood, allowing those large animals to live in what are today frozen arctic regions.

Since my 2011 book, considerable advances have been made in genetic identification of the world's current and ancient populations. As I explained in *Archaeological Evidence*, the clock of genetic distance goes into reverse when isolated population are again mixed. Moreover, genetic distance is calibrated from archaeology and thus itself depends on historical or archaeological identification. Hence, caution is needed when making genetic identification of both modern and ancient peoples. That these studies show Neanderthals breeding with the human population refutes the claim that these cave dwellers were a separate species than humans. Neanderthals were ordinary cave dwelling humans.

There are ways of scientific dating that I can scarcely mention. Dating ancient climate by ocean sediments have replaced layers of peat sediment, but scientists must decide which of countless sites of different layers measure ancient climate. As the case of clouds, rock formations, and tea leaves, one can eventually see what one wishes to see. In 2011, I explained the impossibility of dating by magnetic pole reversals due to the earth's magnetic field as necessarily created by the earth's ionosphere turning with the earth! The theory of plate tectonics was invented to explain the stretching of the ocean basins and the changes in shorelines and elevation of land and mountains due to the Flood and which continue to this day as continents continue to empty their water.

Marine sediment and land peat were impacted by the Flood in a similar way as Flood deposits impacted radiocarbon dating. Likely, annual layers called *varves* were everywhere eroded, disrupted, or buried. But interestingly, the *International Union of Geological Sciences* (IUGS) recently ratified a subdivision of the Holocene. The Upper/Late Holocene Subseries called the *Meghalayan Age* begins 4250 years ago. This most recent epoch of earth history was based on a stalagmite count from the Mawmluh Cave near Meghalaya in northeast India. These latest advances in the geological sciences may have already dated the Flood!

Those inescapable singularities

As retrograde as this paradigm shift may appear to some, it but completes the movement of the sciences to the linear sense of history that has been occurring since the historical view of the

Scriptures challenged Greek philosophy. There was an acceleration of this process in the Scientific Revolution which immediately followed the back-to-the-Bible Reformation and resulted in the birth of the historical sciences.

There have been many attempts to move science back to some eternalist system as happened with the medieval Neoplatonist theology, the Deism of the Enlightenment, the non-historical uniformitarian science of James Hutton and Charles Lyell and which remains the goal of today's multi-world advocates. But despite what popular historians of the geological sciences continue to trumpet, the geological sciences long ago rejected the uniformitarian views of the earth as championed by Hutton and Lyell. The geological catastrophes so despised by Lyell have returned in a big way to the science of geology. Likewise, the Harvard paleontologists Niles Eldredge and the late Stephen J Gould acknowledge that the paleontological record lacks evidence of evolution. Thus, the origin of the basic species points to the kind of singularities that were formerly understood as creation events.

Singularities are extraordinary things that science cannot explain but which must nonetheless be accepted in order that science may usefully function. A *singularity* is but a mathematical term for what might otherwise be called a *miracle,* whether it be a recent empirical discovery or something extraordinary experienced and recorded from ancient times. Explanations by miracles has been opposed on methodological grounds less they hinder the search for natural causes. Indeed, scientists should never introduce speculative miracles for things their own theories cannot explain as was the practice in early modern science and as Isaac Newton did when he suggested that God must on occasion correct the planetary orbits from degradation caused by gravitational attraction between the planets. This gave rise to what has been called the 'God of the gaps.'

But when naturalistic explanations cannot explain things or when attempts to do so have been so unfruitful, we have good reason to recognize the historicity of certain singularities mentioned in the Bible as the creation of the cosmos evidenced by ancient light, together with the expansion of the universe, popularly if incorrectly called the 'Big Bang.' The original creation wasn't the spatial singularity that led to the ad hoc and untestable theory of inflation as now presumed due to this being a time when the ordinary laws of physics breakdown. Rather, as the Scriptures inform us, God began our physical world with his creation of water!

Early modern scientists supposed the water in the creation account concerned only the earth. Isaac Newton taught them to read this account phenomenologically, as would an observer standing on the surface of the earth. But as the text reads, the Creation account concerns both the heavens and the earth. It is not man's but the Creator's perspective. In the light of the biblical narrative, the skies' microwave background radiation can best be understood as evidence of the creation of light on the first day of Creation, while the expansion of the waters on second day of Creation pertains to our expanding universe. Likely, dark matter is but interstellar water, while dark energy comes from God stretching out the heavens.[39] How a universe that is uniformly expanding in all directions can

---

[39] Isaiah 40:22, 42:5, 45:12, 48:13

form galaxies and stars is a matter that the science of cosmogony has yet to adequately explain. Thus, a unique act on the fourth day of Creation was needed to create the galaxies and stars, together with their planets and moons.

The creation of the earth from water prior to the sun, moon, and stars might explain why our planet holds such a privileged place in our physical universe and why the stars of our Milky Way galaxy which were also created from water appear as old as the universe itself. This may explain why plants are also able to grow in the spectrum of infrared light as would have existed from the background radiation still to be observed from every direction of the skies but having been considerably brighter and closer to the spectrum of visible light during the earlier days of our universe. The earth's topsoil created by this vegetation would have eventually washed into the rivers and seas, where after becoming buried to great depths, turns into oil and gas. Such fossil fuels are far older than fossilized plants.

Likewise, it is becoming ever more certain that the miraculous origin of life on earth cannot be explained by natural causes. Due to the absurdly excessive improbability of explaining the origin of life by natural causes, a cause that transcends the physical world is far more probable. When supernatural miracles as those in the sacred scriptures are the simplest, clearest, and most useful understanding, why not accept them? Those who don't will soon be left behind with the old paradigm.

Singularities cannot be explained but they do leave evidence. Indeed, there is plentiful evidence of our world's creation. But if some suppose that our universe might somehow be uncreated or explained by natural causes, not so the survival of Noah's family and all the kinds if animals from a Flood that covered the earth's highest mountains. This explains why, of all the miracles in the Bible, Voltaire found Noah's Flood to be the most difficult to believe. Voltaire hated the very idea of Noah's Flood because it conflicted with his Deist belief that the Creator didn't continue to involve himself in his Creation though such miraculous acts.

Few miracles including those reported in the Bible leave such lasting physical evidence. Even more than the Creation itself, Noah's Flood is the one miracle that could be tested because it signally impacted the dwellings of every human being and the habitats of every animal living upon the earth. We have found the geological and archaeological evidence of this miraculous event to be manifest in the material remains. The result must be world changing.

But neither should we attempt to fully explain such a manifestly miraculous event solely by natural causes as did men like Jean DeLuc, Dolomieu, and William Buckland who believed they had found clear and unequivocal evidence of Noah's Flood. The only miracles that DeLuc allowed for Noah's Flood was the revelation of the coming Flood to Noah, and the divine Providence that directed the Ark to land on the mountain by which we are gathered. That put the impossible task of gathering the animals and protecting them on their journey to and from the Ark entirely on Noah and his family. But the sacred text tells us that the animals were brought to the Ark and dispatched to their present habitats by the same wise and powerful God who created these animals and continues to feed the sparrows even in ordinary times. Our Heavenly Father continues to maintain the earth for the benefit and development of mankind.

The explanatory superiority of historical narratives to scientific theories

The reality of singularities explains the limitation of scientific theories in all the historical sciences and why historical narratives as those of the Bible enlightened the material remains of the past far more than can narratives composed from scientific theories. In the field of archaeology, science may well understand things such as diet, health, domestication of plants and animals, and the crafting of things from stone, clay, metal, and wood, arts that are universal to all human societies. Even so, some ancient crafts such as the precision manufacture of great stone structures remain mysterious. For sure, without historical narratives we can never understand the beliefs, values, concerns, aims, religion, or the identity of those who once occupied archaeological sites. Of course, we were unable even to properly understand the archaeological and geological processes. Without a true historical narrative, we are searching in the dark when we try to understand the material remains of the past.

Science has been perceived and indeed was much responsible for the Industrial Revolution even if social and political factors stemming from religion if less understood were also responsible. Unfortunately, scientific and economic progress brought with it an arrogant tendency to look down on the wisdom of the past. But wisdom comes from experience which belongs to our fathers.

It was due to Charles Lyell's uniformitarian science that the word *science* (or 'men of science') came to replace what was formerly called either *natural philosophy* or *natural theology.* There immediately arose a conflict in which Science opposed the Bible and the Bible was rightly seen as being opposed to Science on account of Lyell's uniformitarian dogma. The conflict was real! Admiration of science used for technology and industry carried over to the historical sciences, though as but new narratives of the past built from theories of the physical sciences.

The physical sciences are uniformitarian because they pertain to passive dead things such as atoms, rocks, rotating planets, steel, machines, and computers. Science may observe but can't explain or create living things where necessarily exists freedom. Freedom is present in living things because all life comes from the sovereign and free God who makes things grow. Thus, materialist science can never explain life nor the appearance of novel orders of life including humans made in the image of the Creator. Scientific narratives such as evolution that claim to explain everything that has or might ever appear cannot be tested. They may only be science fiction as now appears the case.

Having banished the geological evidence of the Flood and having manipulated the evidence in favor of the antiquity of man, Lyell set the stage for his disciple Darwin's theory of evolution by natural selection. Acceptance of the latter empowered the reception of the German higher criticism of the Bible in the English-speaking world that had long resisted these teachings. Higher criticism was *German* because, though beginning in Holland, England, and France, it was largely the product of the needs of German nationalism.

Textual criticism dates from the Renaissance when Lorenzo Valla demonstrated the forgery of the *Donation of Constantine.* It isn't a science but a historical discipline and essential to the history of ideas. Lorenza Valla worked from historical documents. But the aim of the higher criticism was to explain the composition of the Bible according to the kind of political and religious agenda as the German nationalists were manufacturing for their history of the German people. Their manufactured Old Testament historiography was accordingly attributed to four writers or editors abbreviated by the letters J, E, P, and D, none of which dated earlier than Israel's divided kingdoms and presumed no earlier than Judah's Babylonian captivity.

In comparison to higher criticism of the Bible, archaeology is a new science dating only from the nineteenth century. Though prehistoric archaeology has long existed in a crude and formative state, both prehistoric and historical archaeology are based on physical ruins. Not surprisingly, it has been far more favorable to the sacred scriptures than the composition criticism, beginning with the discovery of historical nations and kings mentioned in the Bible, even texts mentioning the Flood in a similar way as in the Bible.

That there exists a connection between the Bible on one hand and ancient texts and archaeology on the other gave new opportunities for speculative skepticism. This initially resulted in the 'Bible versus Babel' debate near the beginning of the twentieth century which was driven by the soon to be discredited Aryan-inspired antisemitism. Composition critics were however happy to suppose, according to the form criticism of Herman Gunkel (1862-1932), that the narratives of the Bible had borrowed from supposed "older" Ancient Near Eastern myths found in these texts. These critics presumed that ancient Israel was a rare nation that lacked its own account of the Flood, thus having to borrow them from their neighbors. But we now have powerful evidence which conforms more to the earliest chapters of Genesis than to the ANE texts.

If the accounts in the Bible, including the mention of ancient gods are based on historical facts, there ought to be many similarities between the Bible and Ancient Near Eastern texts. These ANE texts also depend on archaeological dating. Should the dating that I propose in this monograph be sustained though testing and further research, composition criticism must exit serious consideration as having been but the rationalization of German nationalism together with the arrogant pretentiousness of naturalistic science.

Historical narratives as those of the sacred scriptures should never be classed with myths. In contrast to the imagined and usually strange worlds of myths including the Flood myths of the Ancient Near East, historical narratives found in the sacred scriptures mention specific individuals, nations, events, and places in a world that aside from occasional extraordinary events is the familiar one in which we live. Mention of specific people and events allow a purported historical narrative to be tested, as when I sought to discover whether there existed archaeological evidence of Noah's Flood. The supernatural as reported in the sacred scriptures functions in harmony with the ordinary, the natural. Else a properly designed Ark would not have been necessary to save Noah and the animals on the Ark from drowning.

Building on widespread admiration of science, scientism took control of Western education and no place more than in my America, the world's oldest and most influential secular state. The failure of scientism as became apparent by the middle of the last century gave rise to the postmodern confusion that now dominates Western education, turning science into an instrument of political power but also resulting in social chaos. Because power corrupts, science should never be politicized and never trusted when it has because politicized science is almost certain to be wrong. If religion should not be politicized, neither should science. If the evidence and fruits are so clear, science doesn't need to be politicized.

Why scientists failed to notice the manifest evidence of Noah's Flood

Next to mention is why the many problems created by the fantastic theories upon which the current paradigm is based have gone so unnoticed or unexamined. Why has the fact that the archaeological remains of the prehistoric world consist almost entirely of elaborate burials gotten so little attention? Why haven't the strangeness of the burial practices by which archaeologists explain these burials led to more discussion? Why wasn't the Ice Age hypothesis submitted to examination by civil or mechanical engineers, those who understand solid and fluid mechanics and thus capable of judging the causes and capabilities of ice flow and erosion? Why haven't biblical archaeologists noticed that the hiatus in population following the Early Bronze Age occurs around the biblical date of the Flood? And why did no one recognize that the widespread migration of the third millennium that follows the Early Bronze Age begins at this famous mountain where the Lord God commanded Noah's family and the animals to replenish the earth?

The chief reason is that scholars and scientists who did believe in the historicity of Noah's Flood either presumed that the Flood was a local one or else had destroyed man's archaeological remains. Both young earth and recent old earth Creationists have interpreted the post-Flood world from the archaeological perspective created by Lyell. Thus, the Flood was supposed to have occurred during the sparsely populated Stone Age world of the Ice

Age! Both young and old earth Creationists are much responsible for blinding the eyes of believers to the evidence of Noah's Flood.

Most of the things that I have discussed have been noticed by the best geological and archaeological minds. As in any field, the more experience one gains, the less impressed he is about what he and his peers suppose they know. This includes the historical sciences. Yet, one cannot however solve the many individual problems noted here apart from the wholesale paradigm shift that I have outlined here. Had professional archaeologists and scientists not followed the standard paradigm, prehistory would have become even more muddled as we see in the case of many alternative archaeologies which did sometimes explain the disappearance of ancient civilizations as due to Flood but also explain the rise of these civilizations as due to aliens from space.

Over a long career in technology and science, I discover that the less experience one has in developing technology, the less likely he will be to question what he learns from reading and hearing about technology and science and the more likely he will be charmed by what he learns from the teachers in the historical sciences. By contrast, farmers, small businessmen, entrepreneurs, and especially engineers who are responsible for the safety and success of new kinds of technology and infrastructure are not only more skeptical but are accustomed to checking the basis of all(!) assumptions as well thoroughly testing new theories. Even so, know that science cannot advance without building on the wisdom of the past even if some things must be corrected. Though I have often been astounded at the lack of independent thinking within the historical sciences, I also know that independent thinking is disruptive and likely costly to those wanting to advance their career within existing institutions, whether secular or religious. A cost must be paid to change the world in order to advance truth.

Banishing disenchantment

For those unfamiliar with this philosophical term, *disenchantment* refers to the banishing of spiritual agents from the secular worldview, that which distinguishes the modern from the premodern world. These spirits include a God together with his powerful angels who remain active in history. They also include the preternatural powers of Satan and demons who retain some control over the physical world for the purposes to which God has assigned them. These limited free agents also include our own souls and spirits which for now dwell in physical bodies under control of our spirit when our bodies are healthy and awake, though perhaps other spirits as well. Our bodies also influence our minds and spirits.

Disenchantment includes the banishment of miracles and supernatural revelation, theophanies, and visions because the source of these whether from God or preternatural spirits must necessarily be unseen (i.e. spiritual) powers who aren't subject to the laws governing the physical world. This recognition of spirits is incompatible with naturalistic science that reduces all causes to some physicalist understanding. This is the modern worldview which some equate with the 'Western' or secular way of thinking. But aside from its cultural elite, not even the West has been completely and perhaps even predominantly "Western." Believing there exist more than the physical world has never been incompatible with scientific, technological, and spiritual progress. Rather, the reverse is the case.

The beginnings of disenchantment was the rationalism of Greek philosophy and medieval theology whether pagan, Jewish, Islamic, or Christian. It was however chiefly due to the Protestant desire to challenge the claims of miracles and spiritual powers on the part of Catholics and the revelations of Protestant *Enthusiasts, as* were called those who claimed to have experienced God. Reformed Protestants accomplished this by reducing the acts of God in history to General or Special Providence. The fate of the elect having been predestined, neither the Catholic Church nor Protestant Enthusiasts could effect one's salvation. Providence was invented by the Stoics who attributed supernatural claims to superstition. The term was also used by St. Augustine, which licensed their use by the Reformers. General Providence is essentially the same as science. Special Providence, by which Calvinists explained miracles, allowed Calvinists to affirm the narratives of the Scriptures and occasional Protestant miracles, but encouraged them to explain ever more of the current order of the world according to General Providence, in which case Isaac Newton became a Protestant celebrity.

The elimination of preternatural powers by which the Reformers explained Catholic miracles didn't as much concern Protestants. As Cotton Mather informed those presiding over the Salem Witch Trials, the earliest publications of the Royal Academy, the world's first scientific organization, confirmed the manifest reality of witches. The disenchantment of the world however began shortly after this when Isaac Newton, following the path of Henry More and the great Cambridge polymath Joseph Mede, taught Christians to interpret angels and powers in the books of Daniel and Revelation as referring to historical people and events. For example, angels blowing trumpets were human messengers such as the great pre-

Reformers. The great Red Dragon[40] was the pagan Roman Empire. Newton, who secretly practiced alchemy, reduced witchcraft to the dispensing of drugs. Notwithstanding Newton exorcising witches from the modern world, dispensing of drugs rightly or not continues to cast spells of fear while fears of pandemics and 'witch hunts' beset the era of modern science.

According to Newton, Satan and demons were just lusts of the flesh.[41] Following the Jewish rabbi Maimonides in teaching *mortalism* (the death of the soul until a future resurrection), Newton eliminated the souls in Heaven just as he had eliminated Heavenly angels. The new sciences of sociology and psychology immediately appeared to replace the spiritual and biblical understanding of man. With his disenchantment of the world, Isaac Newton birthed the modern age where the influence of Heaven and the spiritual world are no longer recognized even by many modern Christians. Not everyone, Christian or not, Calvinist or not, and not even John Calvin himself have been completely modern in this reductionist sense.

This reductionist worldview caused our modern world to suppose that such acts of God as storms, earthquakes and climate change can be entirely explained through physical causes. Angry about Americans resistance to accepting Darwin's theory of evolution, Andrew Dickinson White devoted much of *A History of Warfare of*

---

[40] Revelation 12:3

[41] For more on Isaac Newton's lesser known views, c.f. The Newton Project, also Stephen David Snobelen's 'Lust, Pride, and Ambition: Isaac Newton and the Devil'

Science with Theology in Christendom to attacking beliefs that spiritual forces were involved in such occurrences. Due to advances in technology, we now observe these phenomena from afar and accordingly predict their arrival with some accuracy. The weather on the sun and the position of the moon do much determine the earth's climate and weather. But the repeated failures of climate change and short range weather predictions show that we are no closer to understanding why these phenomena occur or when they might occur in the more distant future than could our ancestors.

Having spent much of my early career studying and investigating the phenomena of lightning and the physical processes involved, I am well aware of the importance of human wisdom when it comes to avoiding natural disasters. But I have often enjoyed divine protection. God continues to speak in the storms, just as taught in the Bible. He is alive and active as much in Nature as in the affairs of mankind. I see no reason not to believe that unseen spiritual powers ultimately rule the processes of our physical world which but dimly reflect the larger, greater, richer, and more concrete(!) spiritual reality behind our visible world, things that our physical senses can only dimly perceive.

Humans have a role for tending the earth, though not for the earth's climate which humans didn't create, thus cannot regulate. Unfortunately, money and power corrupt truth. To understand the "consensus science" connected with anything controversial that governments may wish to regulate, one must follow the money or political interest and the impact of the same on determining public opinion. As seen in the impact of man's behavior responsible for the erstwhile climate change discussed here, and greater changes in sea levels since that time than any now feared, man's role in the process may be the opposite of what is currently supposed. Fortunately, we live on an earth which God continues to maintain for the benefit of mankind if not in the way that the Deists had wished. The scriptures also teach us that God uses the earth, the skies, and the weather to to warn, to judge, and to bless mankind.

Peer review that accompanies a paradigm shift

Because humans are fallible, peer review delights any good scientist. That which is authentic doesn't mind being tested because such tests reveal authenticity. As was the case of Einstein, he will himself suggest tests. At the beginning of the second chapter, I explained how this new paradigm was the result of a methodological test of the current belief that there exists no archaeological evidence of Noah's Flood. But however confident we may be in the superiority of the new paradigm of *Scripture Science,* such important new things need to be rigorously examined and tested. Back and forth criticism and response to criticism among peers is how new understandings are established and how science advances.

This requires those with enough understanding of the subjects involved to be able to know whether the important matters have been properly and sufficiently addressed. These are one's peers who write for specialized journals related to the field at hand. The editors of the peer-reviewed journals have considerable influence and power in this regard. Reviews may point to the need for further investigation, especially in areas that may not have been addressed. Reviews by peers admit that claims are at least worthy of review even if only, and whether for good or bad reasons, they have obtained the interest of important segments of the public.

The two fields that most concern the findings presented here are prehistoric archaeology and glacier science. These two fields are no exceptions to the extreme specialization that has overtaken all fields of academia. Though Paleolithic archaeology depends on glacier science and to some extent, glacier science depends on Paleolithic archaeology, few from either scientific field understand the connections as I have discussed here, and thus until now haven't been equipped to examine or question things wrongly believed in either field. This is the unfortunate result of scientific specialization. Archaeological specialization is not only regarding regional focus, but also scientific specialties such as ceramics, digging and publishing methodologies, and dating techniques. Archaeologists working in the same region may use altogether different methodologies plus clashing ways of interpreting the same sites. This is the free inquiry of science.

As do the other historical scientists, archaeologists do often suppose or wish that their "discoveries" should somehow benefit society however academic they may actually be. Unfortunately, many today are content just to write for the tiny few in their own specialized fields. This monograph however suggests that archaeology is potentially a field of great relevance to everyone because it involves a paradigm shift that is world changing.

Peer review that results in major paradigm shifts are of a different order than ordinary peer review. To begin, matters that address the biggest subjects are so multi-disciplinary as to pioneer a radically new field where specialized peers and journals hardly exist. The matter may be so important or may concern what will become a new corpus of knowledge that every educated person or cultural leader will be presumed to know. These are the kind of things that radically change the world.

Examining and changing big picture matters by teaching a new paradigm require engaging culture and institutional leaders in the conversation. The current paradigm arose at a time when pioneering geologists and archaeologists such as Jean DeLuc, Georges Cuvier, Charles Lyell, Louis Agassiz, John Lubbock, V Gordon Childe, Kathlyn Kenyon, and the Leaky family working in Kenya, did write for the educated public, including the leaders of governments. It remains the case that those who most influence the understanding in these fields continue to write for the public or, as in the case of the late Jacob Bronowski or Carl Sagan, communicate through new media such as television and today through social media. Increasingly, those who do so are but journalists and popularizer, especially the writers and producers of documentaries and influential movies who may lack the historical and expert understanding of the things about which they write. Keep in mind however that scientific specialists are also laymen in every scientific field but their own specialties! Forget the expert, we need more generalists for it is the specialists who have kept our world in its latest and greatest Dark Age!

It is of course true that one cannot shift to a new paradigm without the freedom to do so, the kind of thing that institutional commitments keep in check. Fortunately, our current business-driven world makes it possible for individual professionals and even institutions to shift to the new paradigm should there be sufficient public interest and support. As Thomas Kuhn noted, paradigms change when individuals and institutions working under the new paradigm attract more students and funding than those working under the old paradigm. That is already happening as more archaeologists and scientists and soon the whole world become aware of the scientific interests informing the topic of this symposium. Truth is powerful. Those who love truth more than worldly rewards will be the same as those changing the world.

C.S. Peirce (1839-1914), the most erudite and scientifically experienced of all the philosophers of science, explained that theories aren't seriously questioned until their arises a significant reason. The discovery of what appears Noah's Ark at the one tiny spot on our large earth's surface that matches the resting place of the Ark as recorded in our sacred scriptures is such a reason. We now have an opportunity for the kind of public discussion required for a paradigm shift. For hasn't everyone from the most erudite academician and scientist to the smallest child heard of Noah's Ark?

Bringing prosperity to this region and peace to nations by serving as the world center for the science of origins

The last thing concerns the prospects of not only scientists but the entire world soon learning of the most recent common ancestor of every human being and every kind of land animal as once being on this mountain and region. Much can be learned here concerning living and extinct species, things needed for wildlife conservation, that we may learn just how much or how little this should impact land development and wildlife policies.

There is also much to learn from this region about the spread of animal and plant domesticates as needed for the preservation of heritage species. Scientists already know this region is home to the wild ancestors of barley, wheat, grapes, apricots, sheep, and goats. It is likely the home of the domesticated horse and spoked-wheel wagons. The study of architecture, textiles, and other ancient crafts that began in this region will assist anthropologists and the students of crafts from all over the world. Beginning from the scientific center where we are meeting, this region should become the world center for the study of all the historical sciences

At a time when many nations are in trouble, it is helpful for all the peoples of the world to know just how closely we are related. The account of the Flood and Prophet Noah is shared by all three Abrahamic faiths, but memories of the Flood were retained by almost every nation including those living on islands in the Southern Pacific that are most distant from this mountain. A visit to this place should serve as a festive homecoming for all mankind because we are all the descendants of Noah.

I have devoted much attention to the racist understandings of mankind upon which the scientific accounts of mankind's origins due to rejecting the origins of all nations as coming from Noah's family only a little more than four thousand years ago. Those scientific views were advertised as "enlightened." Sadly, they just enveloped the modern world in darkness, excused the mistreatment of peoples, and gave rise to many conflicts and wars, including those which have beset this mountain and region. May the ancient light from this mountain become a beacon for the nations to beat their swords into ploughshares as the world learns again to become a single community as formerly the case of our fathers on this very mountain by which we are gathered.

Annotated Bibliography

The History and Theory of Archaeology

William F. Albright, *From the Stone Age to Christianity* (1957), an erudite summary of archaeological prehistory up to the middle of the twentieth century

V. Gordon Childe, *The Aryans* (1926), *Man Makes Himself* (1951), *New Light on Most Ancient East* (1953), *The Dawn of European Civilization* (1957)

Glyn Daniel, *The Idea of Prehistory (1963), One Hundred Years of Archaeology (1950),* two standard texts for the history of archaeological prehistory

Marvin Harris, *The Rise of Anthropological Theory* (2001)

J.P. Mallory, *In Search of the Indo-Europeans: Language, Archaeology, and Myth* (1989)

Colin Renfrew, *Before Civilization* (1973)

Bruce G. Trigger, *A History of Archaeological Thought (1989)*

David Wilson, *The New Archaeology* (1974)

Archaeological Texts

Amnon Ben-Tor, Editor, *The Archaeology of Ancient Israel* (!992)

Charles Burney, David Lang, *The People of the Hills: Ancient Ararat and Caucasus (1971)*

Michael D. Coe, *The Maya* (1993), *Mexico: from the Olmecs to the Maya* (1993), discusses connections between Asia and the Americas

Harriet Crawford, *Sumer and the Sumerians* (1991)

I.E.S. Edwards et al, Editors *The Cambridge Ancient History* (Vol 1, Part I, 1970)

Robert W. Ehrich, Editor, *Chronologies in Old World Archaeology (Vol I and II, 1992)*

Stuart J. Fiedel, *Prehistory of the Americas,* Second Edition (1992)

James A. Ford, *A Comparison of the Formative Cultures of America* (1969)

Jacquetta Hawkes, *The Atlas of Early Man (1976)*

Kathleen Kenyon, *Digging Up Jericho* (1957)

Samuel Noah Kramer, *The Sumerians* (1963)

Sir John Lubbock, *Pre-Historic Times* (Sixth Edition, 1904), *The Origin of Civilization and the Primitive Condition of Man* (1870)

Betty J. Meggers, Clifford Evans, and Emilio Estrado, *Early Formative Period in America: the Valdiva and Machalio Phases* (1965)

Hans J. Nissen, *The Early History of the Ancient Near East: 9000-2000 BC* (1988)

Konstantine Pitskhelauri, URUK MIGRANTS IN THE CAUCASUS, *Bulletin of the Georgian Academy of Sciences,* vol 6, no. 2, 2012, pp 153-161

Michael Roaf, *Cultural Atlas of Mesopotamia and the Ancient Near East* (1990)
Mitchell S. Rothman, *Uruk, Mesopotamia, and its Neighbors* (2001)
Dean Snow, *The Archaeology of North America* (1976), includes Glacial Kame, Hopewell comparison
Margreet L. Steiner, Ann E. Killibrew, Editors, *The Archaeology of the Levant: c: 8000-332BCE* (2014)
Sir Leonard Woolley, *Excavations at Ur* (1954), Wooley describes his excavation of the Royal Tombs or UR

The History and Theory of Geology

Robert Bakewell, Benjamin Silliman, *An Introduction to Geology* (1839)
Sir John William Dawson, *The Meeting Place of Geology and History* (1894)
Charles Coulston Gillipsie, *Genesis and Geology (1951),* the standard text for how the uniformitarian view triumphed over diluvium and related views
A. Hallam, *Great Geological Controversies,* Second Edition (1989)
Sir Henry H. Howorth, *The Glacial Nightmare and the Flood* (Vol I, 1892; Vol II, 1893), *The Mammoth and the Flood (*1887), *Ice or Water* (Vol I, 1905)
John Imbrie and Katherine Palmer Imbrie, *Ice Ages (1979),* history of Ice Age theory
Charles Lyell, *Principles of Geology* (Volumes I,II, and III, 1830-1833), *The Geological Evidence of the Antiquity of Man* (1863)
Martin J.S. Rudwick, *The Meaning of Fossils* (1976), *Bursting the Limits of Time* (2005), *Men Before Adam* (2008), *Earth's Deep History* (2014), the most comprehensive geological history for the topics covered in the present book

The History of Ideas

Tuska Benes, *In Babel's Shadow: Language, Philology, and the Nation in Nineteenth-Century Germany (2008)*
James E. Force, *William Whiston: Honest Newtonian* (198**5)**
Tom Holland, *Dominion: How the Christian Revolution Remade the World* (2019, Basic Books).
Arthur O. Lovejoy, *The Great Chain of Being* (1964)
Suzanne L. Marchand, *German Orientalism in the Age of Empire: Religion, Race, and Scholarship* (2009), *Down from Olympus: Archaeology and Philhellenism in Germany, 1750-1970* (1996), WHERE DOES HISTORY BEGIN: J.G. HERDER AND THE PROBLEM OF NEAR EASTERN CHRONOLOGY IN THE AGE OF THE ENLIGHTENMENT, *Eighteen-Century Studies,* vol. 47, no. 2, (2014) pp. 157-175
Alister E. McGrath, *Reformation Thought (1993)*
William Paley, *Natural Theology (1802)*

Richard H. Popkin, *The History of Skepticism from Savonoarola to Bayle* (2003), *Isaac la Peyrere (1596-1676) His Life, Work, and Influence* (1987, Brill)
Dorothy Ross, *The Origins of American Social Science* (1991)
Paolo Rossi, T*he Dark Abyss of Time*, translated by Lydia G. Cockrane (1984)
Jonathan Sheehan, *The Enlightenment Bible* (2005)
Stephen D. Snobelen, ISAAC NEWTON, HERETIC, THE STRATEGIES OF A NICODEMITE, *The British Journal for the History of Science*, (December 1999), pp 381-419
Peter Toon, *Puritans, the Millennium, and the Future of Israel* (1970)
James Turner, *Without God, Without Creed: The Origins of Unbelief in America* (1985)

Noah's Ark and Flood

John Warwick Montgomery, *The Quest for Noah's Ark* (1972), covers historical reports of Noah's Ark on Mount Ararat
Ronald L. Numbers, *The Creationists* (1992), the origins of the current versions of young earth Creationism
Philip Ernest Williams, *The Archaeological Evidence of Noah's Flood* (2011)
Davis A. Young, *The Biblical Flood: A Case Study of the Church's Response to Extrabiblical Evidence* (1995)

Quaternary Theory and Science-based Dating

M. J. Aitken, *Science-based Dating in Archaeology* (1990)
Richard B Alley, *The Two-Mile Time Machine: Ice Cores, Abrupt Climate Change and Our Future* (2000), covers attempts to find a chronology in ice cores
M.G.L Baillie, *A Slice Through Time: dendrochronology and precision dating* (1995)
Raymond S. Bradley, *Paleoclimatology: Reconstructing Climates of the Quaternary* (1999)
J.K. Charlesworth, *The Quaternary Era with Special Reference to its Glaciation (Vols I and II,* 1957)
Thomas Cronin, *Principles of Paleoclimatology* (1999)
Richard Foster Flint, *Glacial and Quaternary Geology* (1971)
Stephen Weiner, *Microarchaeology: Beyond the Visible Archaeological Record* (2010)

Made in the USA
Monee, IL
11 November 2026

89eefc2c-386d-4b85-ada3-d43e32ca8f9aR01